CALCULUS FROM THE GROUND UP
SOLUTION GUIDE

Calculus from the Ground Up: Solution Guide

Published in the United States by Blyth Institute Press and BP Learning in Broken Arrow, Oklahoma.

Library of Congress Control Number: 2018945131

ISBN: 978-1-944918-15-6

For author inquiries please send email to info@bplearning.net.

Bookstore bulk order discounts available. Please contact info@bplearning.net for more information.

For more information, please see www.bplearning.net.

2nd printing

Blyth Institute Press

BP
Learning

CALCULUS FROM THE GROUND UP
SOLUTION GUIDE

BY
JONATHAN
BARTLETT

iv

Contents

Chapter 1

Introduction

This is the Solution Guide for the book *Calculus For Everyone* by Jonathan Bartlett. The recommended way to use this guide is as follows:

1. Attempt to solve the question in the book yourself.

2. After solving it, check your answer and your steps.

3. If you could not solve it, read through the solution in this guide once.

4. Now, try to solve it on your own without referring to this guide. Be sure not to skip any steps, even though you know what they are! This is important for helping you to learn how the problem is solved.

5. If you can't solve it without referring to the solution in this guide, then copy the *entire* solution and explanation from the guide, including all of the justifications for each step.

6. After doing this, attempt the problem again without referring to the guide or your notes.

7. Repeat steps until understanding occurs.

8. If you get frustrated after step 5, move on to the next problem. It is better to proceed than to wallow in frustration.

Part I

Preliminaries

4

Chapter 2

A General Method for Solving Mathematics Problems

This chapter had no questions.

Chapter 3

Basic Tools: Lines

1. **Question:** Write out the standard equation for a line (Equation 3.1) five times.

   Solution: $y = mx + b$

2. **Question:** What does m refer to on the standard equation for a line?

   Solution: m refers to the slope of the line.

3. **Question:** What does b refer to on the standard equation for a line?

   Solution: b refers to the y-intercept of the line.

4. **Question:** Draw the line given by the equation $y = 5x - 3$.

   Solution:

   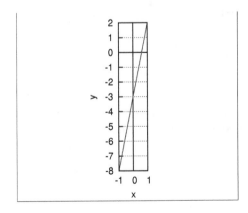

   Explanation: This graph has a slope of 5

and a y-intercept of -3.

5. **Question:** Draw the line given by the equation $y = \frac{1}{3}x + 1$.

   Solution:

   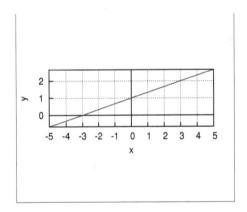

   Explanation: This graph has a slope of $\frac{1}{3}$ and a y-intercept of 1.

6. **Question:** Draw the line given by the equation $y = 2.3x + 4.1$.

   Solution:

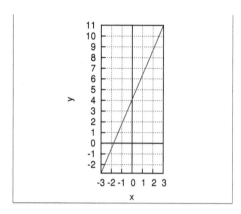

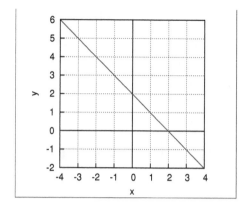

Explanation: This graph has a slope of $\frac{2.3}{1}$ and a y-intercept of 4.1.

7. **Question:** Draw the line given by the equation $5y = 3x + 10$.

Solution:

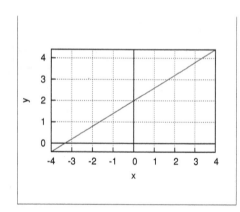

Explanation: Dividing both sides by 5 will put this equation in a standard form: $y = \frac{3}{5}x + 2$. This graph has a slope of $\frac{3}{5}$ and a y-intercept of 2.

8. **Question:** Draw the line given by the equation $y + x = 2$.

Solution:

Explanation: To get this in the standard form for a line, subtract x from both sides, yielding $y = -x + 2$. This equation has a slope of $\frac{-1}{1}$ and a y-intercept of 2.

9. **Question:** Determine the equation of the line given by the point $(2, -1)$ and the slope $\frac{1}{5}$.

Solution: $y = \frac{1}{5}x + \frac{-7}{5}$

Explanation: For this we start out knowing the slope already. So, our equation is $y = \frac{1}{5}x + b$. We are given a point, so, using this point for an x and a y, we can solve for b:

$$y = \frac{1}{5}x + b \qquad \text{the equation}$$

$$-1 = \frac{1}{5}(2) + b \qquad \substack{\text{substituting the } x \text{ and } y \\ \text{from our point}}$$

$$-1 = \frac{2}{5} + b \qquad \text{solve for } b$$

$$-1 - \frac{2}{5} = b$$

$$\frac{-5}{5} - \frac{2}{5} = b$$

$$\frac{-7}{5} = b$$

Since we were given m and we have solved for b, we now know the whole equation:

$$y = \frac{1}{5}x + \frac{-7}{5}$$

10. **Question:** Determine the equation of the line given by the points $(1, 1)$ and $(2, 3)$.

Solution: $y = 2x - 1$

Explanation: For this we can just use Equation 3.6 (or Equation 3.5) directly.

$$y - y_0 = \frac{y_1 - y_0}{x_1 - x_0}(x - x_0) \quad \text{Equation 3.6}$$

$$y - 1 = \frac{3 - 1}{2 - 1}(x - 1) \quad \text{substituting with the two given points}$$

$$y - 1 = \frac{2}{1}(x - 1) \quad \text{simplifying}$$

$$y - 1 = 2(x - 1)$$

$$y - 1 = 2x - 2$$

$$y = 2x - 1 \quad \text{solved}$$

11. **Question:** Determine the equation of the line given by the points $(4, 5)$ and $(-3, 2)$.

    Solution: $y = \frac{3}{7}x + \frac{4}{7}$

    Explanation: This is solved the same way as before.

    $$y - y_0 = \frac{y_1 - y_0}{x_1 - x_0}(x - x_0) \quad \text{Equation 3.6}$$

    $$y - 5 = \frac{2 - 5}{-3 - 4}(x - 4) \quad \text{substituting our particular point values}$$

    $$y - 5 = \frac{-3}{-7}(x - 4) \quad \text{simplifying}$$

    $$y - 5 = \frac{3}{7}x - \frac{12}{7}$$

    $$y = \frac{3}{7}x - \frac{12}{7} + \frac{35}{7}$$

    $$y = \frac{3}{7}x + \frac{23}{7}$$

12. **Question:** Give an analytic equation for determining the equation of a line given a single point (x_0 and y_0) and the slope (m).

    Solution: $y = mx + y_0 - m x_0$

    Explanation: An analytic form of an equation is one where all of the pieces are written out explicitly in mathematical form as an equation.

Here, we are given a slope m, and a point, (x_0, y_0). What we need to do is to show how to build the equation of a particular line out of these points.

The basic line equation is $y = mx + b$. The goal is to come up with an equation that looks like this, but using the pieces we are given. However, the problem is simplified because we are treating the slope (m) as a given. That leaves only b that needs to be solved for. In other words, we need to solve for b (which is not given) using x_0 and y_0 (which is given).

To think about how to do this, remember that (x_0, y_0) is an actual point on the line. That means that the equation holds true when x is x_0 and y is y_0. Therefore, substituting x_0 for x and y_0 for y will still hold the equation true.

Therefore, the equation can be transformed into this:

$$y = m x + b \quad \text{basic line equation}$$

$$y_0 = m x_0 + b \quad \text{using our } x_0 \text{ and } y_0 \text{ for } x \text{ and } y$$

In this equation, the only thing that isn't given from the problem is b. Therefore, we can solve for b (i.e., put b by itself on one side of the equation) by subtracting $m x_0$ from both sides:

$$y_0 = m x_0 + b$$

$$y_0 - m x_0 = b$$

Now we have an equation for b entirely in terms of m, x_0, and y_0. That means that we can take $y_0 - m x_0$ and substitute it for b anywhere where b exists.

So, in our original line equation, we will substitute this for b:

$$y = mx + b \quad \text{basic line equation}$$

$$y = mx + (y_0 - m x_0) \quad \text{substituting in for } b$$

When substituting, it is often helpful to wrap the thing that you are substituing in parentheses as we did above, so that it acts as a unit within the equation (just as b was a single unit in the equation). Then, if it makes sense, you can get rid of the parentheses. Here, because adddition is associative, we can just remove the parentheses, giving us the final answer:

$$y = mx + y_0 - m x_0$$

Chapter 4

Basic Tools: Variables and Functions

1. **Question:** If $f(x) = 3x + 5$, what is $f(7)$?

   Solution: 26

   Explanation: $f(7) = 3 \cdot 7 + 5 = 21 + 5 = 26$

2. **Question:** If $f(b) = b^2 + 2$ what is $f(9)$?

   Solution: 83

   Explanation: $f(9) = 9^2 + 2 = 81 + 2 = 83$

3. **Question:** If $f(x) = 2^x + x$ what is $f(3)$?

   Solution: 11

   Explanation: $f(3) = 2^3 + 3 = 8 + 3 = 11$

4. **Question:** if $f(n) = 2n - 7$ what is $f(q + 5)$

   Solution: $2q + 3$

   Explanation:
   $$
   \begin{aligned}
   f(q + 5) &= 2(q + 5) - 7 \\
   &= 2q + 10 - 7 \\
   &= 2q + 3
   \end{aligned}
   $$

5. **Question:** If $f(x) = x^2 + 3x$, what is $f(z + 1)$?

   Solution: $z^2 + 5z + 4$

   Explanation:
   $$
   \begin{aligned}
   f(z + 1) &= (z + 1)^2 + 3(z + 1) \\
   &= (z^2 + 2z + 1) + (3z + 3) \\
   &= z^2 + 5z + 4
   \end{aligned}
   $$

6. **Question:** If $h(x) = \frac{x^4}{9}$, what is $h(3)$?

   Solution: 9

   Explanation: $h(9) = \frac{3^4}{9} = \frac{81}{9} = 9$

7. **Question:** If $f(x, y) = 3x + 4y$, what is $f(5, 6)$?

   Solution: 39

   Explanation: $f(5, 6) = 3(5) + 4(6) = 15 + 24 = 39$

8. **Question:** If $f(x) = x^2 + 5x$ and $g(x) = 5x - 9$, what is $f(g(x))$ in terms of x?

   Solution: $f(g(x)) = 25x^2 - 65x + 36$

   Explanation:
   $$
   \begin{aligned}
   f(g(x)) &= (5x - 9)^2 + 5(5x - 9) \\
   f(g(x)) &= (25x^2 - 90x + 81) + (25x - 45) \\
   f(g(x)) &= 25x^2 - 65x + 36
   \end{aligned}
   $$

9. **Question:** What is the inverse of the function $f(x) = x + 1$?

Solution: $f^{-1}(x) = x - 1$

Explanation: First, set it equal to y and then solve for x:

$$y = x + 1$$
$$y - 1 = x$$

Then, swap y and x:

$$x - 1 = y$$

Now write in the form of an inverse function:

$$f^{-1}(x) = x - 1$$

10. **Question:** What is the inverse of the function $f(x) = 5x$?

Solution: $f^{-1}(x) = \frac{1}{5}x$

Explanation:

$y = 5x$ written as an equation

$\frac{1}{5}y = x$ solve for x

$\frac{1}{5}x = y$ swap variable names

$f^{-1}(x) = \frac{1}{5}x$ write in inverse function notation

11. **Question:** What is the inverse of the function $g(x) = \frac{x}{5}$?

Solution: $g^{-1}(x) = 5x$

Explanation: Start out by just using a variable for $g(x)$: $y = \frac{x}{5}$. Next, solve for x:

$$y = \frac{x}{5}$$
$$5y = x$$

Now, swap your x and y: $y = 5x$. So the inverse function is $g^{-1}(x) = 5x$.

12. **Question:** What is the inverse of the function $g(x) = (x + 1)^3$?

Solution: $g^{-1}(x) = \sqrt[3]{x} - 1$

Explanation: We will take the original equation and solve for x:

$$y = (x + 1)^3$$
$$\sqrt[3]{y} = x + 1$$
$$\sqrt[3]{y} - 1 = x$$

Now we will swap x and y, giving us $y = \sqrt[3]{x} - 1$. Therefore, the inverse function is $g^{-1}(x) = \sqrt[3]{x} - 1$.

13. **Question:** What is the inverse of the function $f(x) = \ln(x^2)$?

Solution: $f^{-1}(x) = \sqrt{e^x}$

Explanation: We will take the original equation and solve for x:

$$y = \ln(x^2)$$
$$e^y = e^{\ln(x^2)}$$
$$e^y = x^2$$
$$\sqrt{e^y} = x$$

If we swap x and y, we can then say that $f^{-1}(x) = \sqrt{e^x}$

14. **Question:**   Convert the implicit function $xy = x + 1$ into an explicit function of x.

Solution: $y = 1 + \frac{1}{x}$

Explanation: Converting to an explicit function of x means getting the y by itself on one side of the equation so that y can be determined just from x:

$$xy = x + 1$$
$$y = \frac{x + 1}{x}$$
$$y = 1 + \frac{1}{x}$$

15. **Question:** Convert the implicit function $x + y = 3y$ into an explicit function of x.

Solution: $y = \frac{x}{2}$

Explanation: Converting to an explicit function of x means getting the y by itself on one side of the equation so that y can be determined just from x:

$$x + y = 3y$$
$$x = 3y - y$$
$$x = 2y$$
$$\frac{x}{2} = y$$
$$y = \frac{x}{2}$$

16. **Question:** Convert the implicit function $x + y = 3y$ into an explicit function of y.

Solution: $x = 2y$

Explanation: Here, we are looking for a function of y, so that means that x has to be all by itself:

$$x + y = 3y$$
$$x = 3y - y$$
$$x = 2y$$

17. **Question:** Solve the following equation for x: $\tan^2(x) + 6\tan(x) + 3 = 0$. Use radians when doing calculations.

Solution: x can be -0.5032 or -1.3893

Explanation: To solve this, you need to recognize that this looks a lot like a quadratic equation. In fact, it *is* a quadratic equation if you think of $tan(x)$ as a separate function. We can replace this function with a new variable which we will define to be a function of x. We will call this new variable u, and therefore

$u = tan(x)$. Therefore, we can replace $tan(x)$ with u wherever we want:

$$\tan^2(x) + 6\tan(x) + 3 = 0 \quad \text{original equation}$$
$$u^2 + 6u + 3 = 0 \quad \text{substituting } u$$

$$u = \frac{-b \pm \sqrt{b^2 - 4ac}}{2a} \qquad \text{quadratic formula}$$
$$u = \frac{-6 \pm \sqrt{6^2 - 4 \cdot 1 \cdot 3}}{2 \cdot 1}$$
$$u = \frac{-6 \pm \sqrt{36 - 12}}{2}$$
$$u = \frac{-6 \pm \sqrt{24}}{2}$$
$$u \approx \frac{-6 \pm 4.899}{2}$$
$$u \approx -0.5505 \text{ or } -5.4495$$

That gives us answers for u, but we are really trying to solve for x. Therefore, you need to find the inverse function for u to get x back out:

$$u = \tan(x)$$
$$x = \arctan(u)$$

Therefore, using the values we have received for u, we can solve for x:

$$x \approx \arctan(-0.5505)$$
$$\approx -0.5032$$

For our other result:

$$x \approx \arctan(-5.4495)$$
$$\approx -1.3893$$

Therefore, the values for x can be -0.5032 or -1.3893.

18. **Question:** If $f(x) = 3x$, $g(x) = x^2$ and $h(x) = f(g(x))$, what is $h(x)$ in terms of x?

Solution: $h(x) = 3x^2$

Explanation: To work this, you start replacing from the innermost functions:

$h(x) = f(g(x))$ Original equation

$h(x) = f(x^2)$ Replacing $g(x)$ with its definition

$h(x) = 3(x^2)$ Replacing $f(x)$ with its definition

19. **Question:** What is the value of $h(5)$ in the previous exercise?

Solution: 75

Explanation:

$$h(x) = 3x^2$$
$$h(5) = 3 \cdot 5^2$$
$$h(5) = 3 \cdot 25$$
$$h(5) = 75$$

20. **Question:** If $f(x) = \sin(x) + 4$, $g(x) = 2\cos(x) - 6$, and $h(x) = f(x) + g(x)$, what is $h(x)$ in terms of x?

Solution: $h(x) = \sin(x) + 2\cos(x) - 2$

Explanation: To solve this, we simply replace the functions with what they represent:

$$h(x) = f(x) + g(x) \qquad \text{original}$$
$$h(x) = (\sin(x) + 4) + (2\cos(x) - 6) \qquad \text{substituting}$$
$$h(x) = \sin(x) + 2\cos(6) - 2 \qquad \text{simplifying}$$

21. **Question:** Take the equation $y = 2x + 3$. Let's say we have a function $f(x) = 2x$. Rewrite the equation using $f(x)$.

Solution: $y = f(x) + 3$

Explanation: To use this, we simply replace occurrences of $2x$ with $f(x)$:

$$y = 2x + 3$$
$$y = f(x) + 3$$

22. **Question:** Take the equation $y = 6x + 5$. Let's say we have a function $f(x) = 3x$. Rewrite the equation using $f(x)$.

Solution: $y = 2f(x) + 5$

Explanation: To solve this problem, you have to realize that $6x$ is the same as $2 \cdot 3x$ so that you can substitute $f(x)$ for it:

$$y = 6x + 5$$
$$y = 2 \cdot 3x + 5$$
$$y = 2f(x) + 5$$

23. **Question:** Take the equation $y = 2x$. Let's say we have a function $f(x) = x - 1$. Rewrite the equation using $f(x)$.

Solution: $y = 2f(x) + 2$

Explanation: In this case, there is nothing that we can *directly* substitute for $f(x)$. In order to do this, we have to imagine what the equation would need to have in order to make this substitution. We can *manipulate* the equation so that it does in fact give us something we can substitute for.

We want an instance of $x-1$ in the equation so we can substitute in our function. The easiest way to do this is to start by getting x all by itself:

$$y = 2x$$
$$\frac{y}{2} = x$$

Now that x is by itself, we can substract 1 from both sides of the equation:

$$\frac{y}{2} = x$$
$$\frac{y}{2} - 1 = x - 1$$

Now we have our function of x! So we can say:

$$\frac{y}{2} - 1 = f(x)$$

Now, we want to put the equation back in order.

So, we will re-solve for y:

$$\frac{y}{2} - 1 = f(x)$$

$$\frac{y}{2} = f(x) + 1$$

$$y = 2f(x) + 2$$

Chapter 5

Basic Tools: Graphs

1. **Question:** What is the most number of humps that the following polynomial can have?
$y = x^4 - 10x^3 + 35x^2 - 50x + 24$

   Solution: 3

   Explanation: The polynomial is of degree 4, therefore the maximum number of humps is one less.

2. **Question:** What is the most number of humps that the following polynomial can have?
$y = 49x^3 - 4x^6 - 14x^5 + 56x^4 - 196x^2 - 36x + 144 + x^7$

   Solution: 6

   Explanation: The polynomial is of degree 7, therefore the maximum number of humps is one less. Notice that the x^7 is not the first term, but the order of the terms doesn't matter.

3. **Question:** What is the most number of humps that the following polynomial can have?
$y = 3x + 4$

   Solution: 0

   Explanation: The polynomial is of degree 1, therefore it cannot have any humps. x is another way of saying x^1.

4. **Question:** What is the equation of the following graph if shifted to the left five units and down one?
$y = x^2 + 2x - 3$

   Solution: $(y + 1) = (x + 5)^2 + 2(x + 5) - 3$
or simplified as $y = x^2 + 12x + 31$

   Explanation: To shift it down one unit (i.e., -1), all occurrences of y are replaced by $(y + 1)$. To shift it to the left five units (i.e., -5), all occurrences of x are replaced by $(x + 5)$.

5. **Question:** What is the equation of the following graph if shifted up two units and to the left three units?
$5y = x \sin(x)$

   Solution: $5(y - 2) = (x + 3) \sin(x + 3)$

   Explanation: To shift it up two units (i.e., $+2$), all occurrences of y are replaced by $(y - 2)$. To shift it to the left three units (i.e., -3), all occurrences of x are replaced by $(x + 3)$.

6. **Question:** Draw the graph of $y = 5x$ on graph paper. Now draw the inverse of this function.

   Solution: $y = 5x$

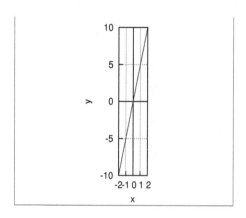

Inverse:

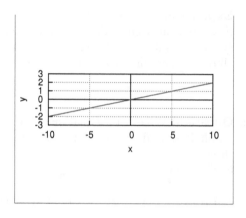

Explanation: The second drawing is a reflection over the line $y = x$ of the first graph.

7. **Question:** Draw the graph of $y = x^2$ on graph paper. Now draw the inverse of this function.

Solution: $y = x^2$

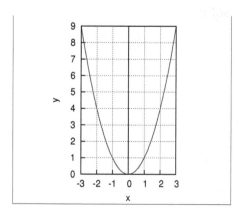

Inverse:

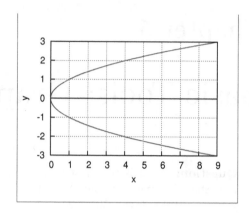

Explanation: The second drawing is a reflection over the line $y = x$ of the first graph. Note that this is not a true function. If only half of the graph is rendered, that would also be correct, as that would actually be a true function.

8. **Question:**   Draw graphs of the following functions from $x = -6$ to $x = 6$: $\sin(x)$, $\cos(x)$, and $\ln(x)$. Draw 2^x from $-3 \leq x \leq 3$.

Solution: $y = \sin(x)$

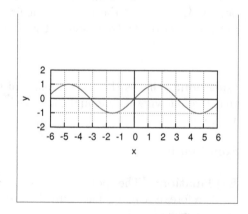

$y = \cos(x)$

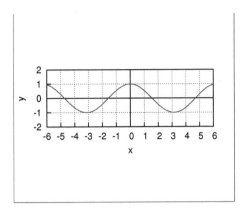

$$y = \ln(x)$$

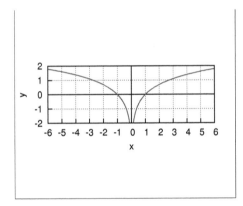

$$y = 2^x$$

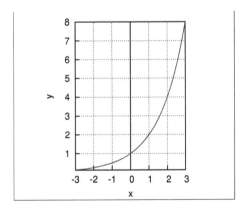

Chapter 6

Calculus: The Big Picture

1. **Question:** Describe the integral in your own words.

   Solution: An integral is an equation that gives you the area under the curve formed by another equation.

2. **Question:** Describe the derivative in your own words.

   Solution: A derivative is an equation that gives you the slope of a curve formed by another equation.

3. **Question:** Think about the graph of $y = 6$. What is the area under the graph at $x = 3$ (i.e., the area bounded by the x-axis, y-axis, the actual graph $y = 6$, and the vertical line at the endpoint, $x = 3$).

   Solution: 18

   Explanation: The graph of $y = 6$ is a straight line. When combined with the two axes and another vertical line, it will produce a rectangle. The height will be 6 and the width will be whatever x value we draw the line at. The area of a rectangle is *width · height*, therefore, for $x = 3$, the area is $6 \cdot 3 = 18$.

4. **Question:** Think about the graph of $y = 3x$. What is the area under the graph at $x = 2$?

   Solution: 6

   Explanation: The graph of $y = 3x$ forms

a line. When combined with the x axis and another vertical line, it will produce a triangle. The height will be $3x$ and the width will be x. The area of a triangle is $\frac{1}{2}base \cdot height$. Therefore, the area will be $\frac{1}{2}3x \cdot x$ or $\frac{3}{2}x^2$. At $x = 2$ this becomes $\frac{3}{2}2^2 = 6$.

5. **Question:** Given the equation of the line $y = 5x + 3$, what is the slope of the line? Since that is the slope for every value of x, what would the *equation* of the derivative be?

   Solution: The slope is 5. The equation of the derivative is therefore $y' = 5$.

6. **Question:** Plot the equation $y = 2x^2 - 3$ on a sheet of paper from $x = -1$ to $x = 3$. Use a ruler to estimate the slope of this equation at $x = 2$.

   Solution: The equation generates the following graph with the given tangen line drawn:

   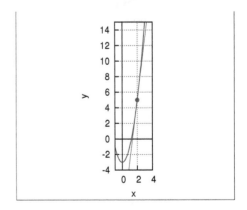

   At $x = 2$ the slope is 8. The drawn slope *does not* need to be anywhere near exact. Based on inconsistencies in the way people draw, any positive slope is an acceptable answer. The most important thing is that the slope is estimated using a line that is as close to the tangent as possible to the curve drawn, and that the listed slope be calculated from the tangent line drawin.

7. **Question:** Plot the equation of $y = x^3 - x^2$ on a sheet of paper from $x = 0$ to $x = 2$ (plot every 0.5 increment on the x axis). Calculate an estimation of the value of the slope at $x = 1$ by finding the slope between the point on the curve at $x = 1$ and $x = 1.1$. Calculate the slope again between the point at $x = 1$ and $x = 1.01$. What do you think the "real" slope actually is?

Solution: The equation generates the following graph:

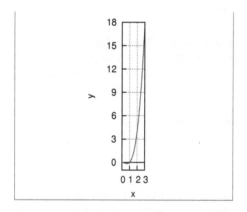

The slope from 1 to 1.1 is 1.21. The slope from 1 to 1.01 is 1.0201. The real slope at this point is 1.

Explanation: To calculate the slope between the two close points, we will calculate the y values for those points and use the point-slope formula to find the slope. For $x = 1$:

$$y = x^3 - x^2$$
$$= 1^3 - 1^2$$
$$= 1 - 1$$
$$= 0$$

For $x = 1.1$:

$$y = x^3 - x^2$$
$$= (1.1)^3 - (1.1)^2$$
$$= 1.331 - 1.21 \qquad = 0.121$$

So, the first point is at $(1, 0)$ and the second point is at $(1.1, 0.121)$. Using these points, we can use the point-slope formula to calculate the slope between these points:

$$m = \frac{y_1 - y_0}{x_1 - x_0} \qquad \text{point-slope formula}$$
$$= \frac{0.121 - 0}{1.1 - 1} \qquad \text{using our data points}$$
$$= \frac{0.121}{0.1} \qquad \text{simplifying}$$
$$= 1.21$$

Therefore, the slope between those two points is 1.21.

Now, for the other two points, we need to find out the y value where $x = 1.01$:

$$y = x^3 - x^2$$
$$= 1.01^3 - 1.01^2$$
$$= 1.030301 - 1.0201$$
$$= 0.010201$$

Now we can use point-slope again:

$$m = \frac{y_1 - y_0}{x_1 - x_0} \qquad \text{point-slope formula}$$
$$= \frac{0.010201 - 0}{1.01 - 1} \qquad \text{using our data points}$$
$$= \frac{0.010201}{0.01} \qquad \text{simplifying}$$
$$= 1.0201$$

8. **Question:** Take the equation $y = x^2 + 2x$ and calculate an estimation of the slope at $x = 3$ by finding the slope between the point on the curve at $x = 3$ and $x = 3.1$. Calculate the slope again between the point at $x = 3$ and $x = 3.01$.

Solution: The first two points are $(3, 15)$ and $(3.1, 15.81)$ with a slope of 8.1 between the points. The second two points are $(3, 15)$ and $(3.01, 15.0801)$ with a slope of 8.01 between them.

Explanation: To calculate the slope between the two points, we will calculate the y values for those points and use the point-slope formula to find the slope. For $x = 3$:

$$y = x^2 + 2x$$
$$= (3)^2 + 2(3)$$
$$= 9 + 6$$
$$= 15$$

For $x = 3.1$:

$$y = x^2 + 2x$$
$$= (3.1)^2 + 2(3.1)$$
$$= 9.61 + 6.2$$
$$= 15.81$$

So, the first point is at $(3, 15)$ and the second point is at $(3.1, 15.81)$. We can use the point-slope formula to calculate the slope between these points:

$$m = \frac{y_1 - y_0}{x_1 - x_0} \quad \text{point-slope formula}$$
$$= \frac{15.81 - 15}{3.1 - 3} \quad \text{using our data}$$
$$= \frac{0.81}{0.1} \quad \text{simplifying} \qquad = 8.1$$

Therefore, the slope between these two points is 8.1. This is our first estimate for the slope at $x = 3$.

For our second estimate, the second point will be closer. The first point will be the same $(3, 15)$, but we will use $x = 3.01$ for our second point. So, first, we need to find the y value for $x = 3.01$:

$$y = x^2 + 2x$$
$$= (3.01)^2 + 2(3.01)$$
$$= 9.0601 + 6.02$$
$$= 15.0801$$

So our second point is $(3.01, 15.0801)$. Now, we can use the point-slope formula again to find the slope between $(3, 15)$ and $(3.01, 15.0801)$:

$$m = \frac{y_1 - y_0}{x_1 - x_0} \quad \text{point-slope formula}$$
$$= \frac{15.0801 - 15}{3.01 - 3} \quad \text{our data points}$$
$$= \frac{0.0801}{0.01} \quad \text{simplifying}$$
$$= 8.01$$

Therefore, the slope between these two points is 8.01.

9. **Question:** Take the equation $y = 3x^2 - x$ and calculate an estimation of the slope at $x = 4$. In order to estimate the slope, for the second point use an x value that is 0.001 away from your real x value.

Solution: The slope here is approximately 24.003.

Explanation: To estimate the slope at $x = 4$ we need two points. The first point will be the point where $x = 4$. To get this point, we need the y value for $x = 4$:

$$y = 3x^2 - x$$
$$= 3(4)^2 - (4)$$
$$= 3 \cdot 16 - 4$$
$$= 48 - 4$$
$$= 44$$

So, the first point will be $(4, 44)$. The second point will be 0.001 away from the first point's x value, so its x value will be $x = 4.001$. Now we need to calculate the y value:

$$y = 3x^2 - x$$
$$= 3(4.001)^2 - 4.001$$
$$= 3 \cdot 16.008001 - 4.001$$
$$= 48.023003$$

Therefore, the second point is $(4.001, 48.023003)$. The point-slope formula will tell us the slope:

$$m = \frac{y_1 - y_0}{x_1 - x_0} \quad \text{point-slope formula}$$
$$= \frac{48.023003 - 48}{4.001 - 4} \quad \text{our data}$$
$$= \frac{0.023003}{0.001} \quad \text{simplifying}$$
$$= 23.003$$

Therefore, the slope at $x = 4$ is approximately 23.003.

Part II

The Derivative

Chapter 7

The Derivative

1. **Question:** Try to derive Equation 7.4 starting with the point-slope formula for yourself. Show the steps.

   Solution: Students are not required to show precise steps. The important part is for them to think about the slope formula and where it comes from. Most students will probably not be this explicit in listing the steps.

$m = \dfrac{y_1 - y_0}{x_1 - x_0}$	point-slope formula
$x_0 = x$	the first point will have x as its x value
$y_0 = f(x)$	the first point's y value can be found by applying f to x
$x_1 = x + h$	the second point's x value is beyond x by an arbitrarily small amount which we will call h
$y_1 = f(x + h)$	we get the second point's y value by putting in x_1 into the function f
$m = \dfrac{f(x + h) - f(x)}{(x + h) - (x)}$	substituting in our particular points
$m = \dfrac{f(x + h) - f(x)}{h}$	simplifying the denominator
$y' = m$	the derivative is the slope
$y' = \dfrac{f(x + h) - f(x)}{h}$	equation of the derivative

2. **Question:** Write Equation 7.4 five times.

   Solution:

   $$y' = \frac{f(x + h) - f(x)}{h}$$

3. **Question:** Use Equation 7.4 to find the derivative of $y = x^2$.

   Solution: $y' = 2x$

   Explanation:

   $$f(x) = x^2$$
   $$y' = \frac{f(x + h) - f(x)}{h}$$
   $$= \frac{(x + h)^2 - x^2}{h}$$
   $$= \frac{x^2 + 2hx + h^2 - x^2}{h}$$
   $$= \frac{2hx + h^2}{h}$$
   $$= 2x + h$$
   $$= 2x + 0$$
   $$y' = 2x$$

4. **Question:** Use Equation 7.4 to find the derivative of $y = 3x^2$.

   Solution: $y' = 6x$

Explanation:

$$f(x) = 3x^2$$

$$y' = \frac{f(x+h) - f(x)}{h}$$

$$= \frac{3(x+h)^2 - 3x^2}{h}$$

$$= \frac{3(x^2 + 2hx + h^2) - 3x^2}{h}$$

$$= \frac{3x^2 + 6hx + 3h^2 - 3x^2}{h}$$

$$= \frac{6hx + 3h^2}{h}$$

$$= 6x + 3h$$

$$= 6x + 3 \cdot 0$$

$$y' = 6x$$

5. **Question:** Find the derivative of $y = 3x^2 + 3$. Compare this to the result you got for the previous answer.

   Solution: $y' = 6x$

   Explanation:

$$f(x) = 3x^2 + 3$$

$$y' = \frac{f(x+h) - f(x)}{h}$$

$$= \frac{(3(x+h)^2 + 3) - (3x^2 + 3)}{h}$$

$$= \frac{(3(x^2 + 2hx + h^2) + 3) - (3x^2 + 3)}{h}$$

$$= \frac{3x^2 + 6hx + 3h^2 + 3 - 3x^2 - 3}{h}$$

$$= \frac{6hx + 3h^2}{h}$$

$$= 6x + 3h$$

$$= 6x + 3 \cdot 0$$

$$y' = 6x$$

This is the same result as for the previous equation (note that if you look at the graphs of both equations, you will see that they always have the same slope even though they have different y values).

6. **Question:** Find the derivative of $y = 6x^2$.

   Solution: $y' = 12x$

   Explanation:

$$f(x) = 6x^2$$

$$y' = \frac{f(x+h) - f(x)}{h}$$

$$= \frac{6(x+h)^2 - 6x^2}{h}$$

$$= \frac{6(x^2 + 2hx + h^2) - 6x^2}{h}$$

$$= \frac{6x^2 + 12hx + 6h^2 - 6x^2}{h}$$

$$= \frac{12hx + 6h^2}{h}$$

$$= 12x + 6h$$

$$= 12x + 6 \cdot 0$$

$$y' = 12x$$

7. **Question:** Find the derivative of $y = 4x^3$.

   Solution: $y' = 12x^2$

   Explanation:

$$f(x) = 4x^3$$

$$y' = \frac{f(x+h) - f(x)}{h}$$

$$= \frac{4(x+h)^3 - 4x^3}{h}$$

$$= \frac{4(x^3 + 3hx^2 + 3h^2x + h^3) - 4x^3}{h}$$

$$= \frac{4x^3 + 12hx^2 + 12h^2x + 4h^3 - 4x^3}{h}$$

$$= \frac{12hx^2 + 12h^2x + 4h^3}{h}$$

$$= 12x^2 + 12hx + 4h^2$$

$$= 12x^2 + 12 \cdot 0 \cdot x + 4 \cdot 0^2$$

$$y' = 12x^2$$

Note that this is the same solution as the prior question.

8. **Question:** Find the derivative of $y = x^2+2x+3$.

   Solution: $y' = 2x + 2$

   Explanation:

   $f(x) = x^2 + 2x + 3$

   $$y' = \frac{f(x+h) - f(x)}{h}$$

   $$= \frac{((x+h)^2 + 2(x+h) + 3) - (x^2 + 2x + 3)}{h}$$

   $$= \frac{(x^2 + 2hx + h^2 + 2x + 2h + 3) - (x^2 + 2x + 3)}{h}$$

   $$= \frac{x^2 + 2hx + h^2 + 2x + 2h + 3 - x^2 - 2x - 3}{h}$$

   $$= \frac{2hx + h^2 + 2h}{h}$$

   $$= 2x + h + 2$$

   $$= 2x + 0 + 2$$

   $$y' = 2x + 2$$

9. **Question:** Find the derivative of $y = x^2+2x+5$.

   Solution: $y' = 2x + 2$

   Explanation:

   $f(x) = x^2 + 2x + 5$

   $$y' = \frac{f(x+h) - f(x)}{h}$$

   $$= \frac{((x+h)^2 + 2(x+h) + 5) - (x^2 + 2x + 5)}{h}$$

   $$= \frac{(x^2 + 2hx + h^2 + 2x + 2h + 5) - (x^2 + 2x + 5)}{h}$$

   $$= \frac{x^2 + 2hx + h^2 + 2x + 2h + 5 - x^2 - 2x - 5}{h}$$

   $$= \frac{2hx + h^2 + 2h}{h}$$

   $$= 2x + h + 2$$

   $$= 2x + 0 + 2$$

   $$y' = 2x + 2$$

Chapter 8

Basic Derivative Rules for Polynomials

1. **Question:** Find the derivative of $y = x^{27}$.

   Solution: $y' = 27x^{26}$

   Explanation: The power rule

2. **Question:** Find the derivative of $y = x^9$.

   Solution: $y' = 9x^8$

   Explanation: The power rule

3. **Question:** Find the derivative of $y = 7x^5$.

   Solution: $y' = 35x^4$

   Explanation:

   $y = 7x^5$

   $y' = 7 \cdot 5x^4$ power and constant multipler rules

   $y' = 35x^4$ simplifying

4. **Question:** Find the derivative of $y = 8x^2$.

   Solution: $y' = 16x$

   Explanation:

   $y = 8x^2$

   $y' = 8 \cdot 2x^1$ power and constant multiplier rules

   $y' = 16x$ simplifying

5. **Question:** Find the derivative of $y = x^{13}$.

   Solution: $y' = 13x^{12}$

   Explanation: The power rule

6. **Question:** Find the derivative of $y = 2x$.

   Solution: $y' = 2$

   Explanation: The constant multiplier rule

7. **Question:** Find the derivative of $y = 373$.

   Solution: $y' = 0$

   Explanation: The constant rule

8. **Question:** Find the derivative of $y = 6^x$.

   Solution: $y' = \ln(6)6^x$

   Explanation: The exponent rule

9. **Question:** Find the derivative of $y = 25x^{27}$.

   Solution: $y' = 675x^{26}$

   Explanation: Power rule and constant

31

multiplier rule:

$$y = 25x^2 7$$
$$y' = 27 \cdot 25x^{26}$$
$$y' = 675x^{26}$$

10. **Question:** Find the derivative of $y = 2\sqrt{x}$

Solution: $y' = \frac{1}{\sqrt{x}}$

Explanation:

$$y = 2\sqrt{x}$$
$$y = 2x^{\frac{1}{2}} \qquad \text{convert to exponent}$$
$$y' = 2 \cdot \frac{1}{2}x^{-\frac{1}{2}} \qquad \text{exponent rule}$$
$$y' = x^{-\frac{1}{2}} \qquad \text{simplify}$$
$$y' = \frac{1}{\sqrt{x}}$$

11. **Question:** Find the derivative of $y = 4^x - 23x$.

Solution: $y' = \ln(4)4^x - 23$

Explanation:

$$y = 4^x - 23x$$
$$f(x) = g(x) - h(x)$$
$$f'(x) = g'(x) - h'(x) \quad \text{addition rule}$$
$$g(x) = 4^x$$
$$g'(x) = \ln(4)4^x \qquad \text{exponent rule}$$
$$h(x) = 23x$$
$$h'(x) = 23 \qquad \text{constant multiple rule}$$
$$f'(x) = \ln(4)4^x - 23$$
$$y' = \ln(4)4^x - 23$$

12. **Question:** Find the derivative of $y = x^2 + x$.

Solution: $y' = 2x + 1$

Explanation: Power rule, constant multiple rule, and addition rule

13. **Question:** Find the derivative of $y = \frac{5}{\sqrt{x}}$

Solution: $y' = \frac{-5}{2}x^{-\frac{3}{2}}$

Explanation:

$$y = \frac{5}{\sqrt{x}}$$
$$y = 5x^{-\frac{1}{2}} \qquad \text{rewrite as exponent}$$
$$y' = 5 \cdot \frac{-1}{2}x^{-\frac{3}{2}} \qquad \text{power rule and constant multiplier rule}$$
$$y' = \frac{-5}{2}x^{-\frac{3}{2}} \qquad \text{simplifying}$$

14. **Question:** Find the derivative of $y = 3x^2 + 2x + 5$.

Solution: $y' = 6x + 2$

Explanation:

$$y = 3x^2 + 2x + 5$$
$$y' = 3 \cdot 2x^1 + 2 + 0$$
$$y' = 6x + 2$$

15. **Question:** Find the derivative of $y = 25e^x + 23 \cdot 4^x - x^3$.

Solution: $y' = 25e^x + 23\ln(4)\,4^x - 3x^2$

Explanation:

$$y = 25e^x + 23 \cdot 4^x - x^3$$

$f(x) = a(x) + b(x) - c(x)$	addition rule
$f'(x) = a(x) + b(x) - c(x)$	
$a(x) = 25e^x$	
$a'(x) = 25e^x$	exponent rule and constant multiplier rule
$b(x) = 23\,4^x$	
$b'(x) = 23\,\ln(4)\,4^x$	exponent rule and constant multiplier rule
$c(x) = x^3$	
$c'(x) = 3x^2$	power rule

$$f'(x) = 25e^x + 23\,\ln(4)\,4^x - 3x^2$$

$$y' = 25e^x + 23\,\ln(4)\,4^x - 3x^2$$

16. **Question:** Find the derivative of $y = x^{\frac{2}{5}} + \frac{6}{x^2} - \frac{2}{\sqrt{x}}$

Solution: $y' = \frac{2}{5}x^{\frac{-3}{5}} - 12x^{-3} + x^{\frac{-3}{2}}$

Explanation:

$$y = x^{\frac{2}{5}} + \frac{6}{x^2} - \frac{2}{\sqrt{x}}$$

$y = x^{\frac{2}{5}} + 6x^{-2} - 2x^{\frac{-1}{2}}$	convert to ordinary powers

$$y' = \frac{2}{5}x^{\frac{-3}{5}} + -2 \cdot 6x^{-3} - 2 \cdot \frac{-1}{2}x^{\frac{-3}{2}} \quad \text{apply derivative rules}$$

$$y' = \frac{2}{5}x^{\frac{-3}{5}} - 12x^{-3} + x^{\frac{-3}{2}}$$

17. **Question:** Find the derivative of $y = (x^2 + 3x)(x - 1)$.

Solution: $y' = 3x^2 + 4x - 3$

Explanation: You cannot find the derivative of this directly, since the multiplication is by a function, not a constant. Therefore, we need to apply the multiplication first, and then do the derivative.

$$y = (x^2 + 3x)(x - 1)$$

$$y = x^3 + 3x^2 - x^2 - 3x$$

$$y = x^3 + 2x^2 - 3x$$

$$y' = 3x^2 + 4x - 3$$

18. **Question:** For the equation $y = x^3 - x^2 + 5x$, what is the slope of that equation at $x = 10$?

Solution: The derivative of the equation is $y' = 3x^2 - 2x + 5$, and therefore the slope of the original equation at $x = 10$ is 285.

Explanation: The derivative of an equation tells you the slope at any given x value. Therefore, we need to first find the derivative:

$$y = x^3 - x^2 + 5x$$

$$y' = 3x^2 - 2x + 5$$

Now that we have the derivative, to find the slope of the original equation at $x = 10$, we just use 10 as our x value in the derivative:

$$y' = 3x^2 - 2x + 5$$

$$= 3(10)^2 - 2(10) + 5$$

$$= 3 \cdot 100 - 20 + 5$$

$$= 285$$

The slope of the original equation at $x = 10$ is 285.

19. **Question:** For the equation $y = e^x + x^3 - 5x$, find the slope at $x = 0$.

Solution: The derivative of the equation is $y' = e^x + 3x^2 - 5$ and therefore the slope of the original equation at $x = 0$ is -4.

Explanation: The derivative of an equation tells you the slope at any given x value. Therefore, we need to first find the derivative:

$$y = e^x + x^3 - 5x$$

$$y' = e^x + 3x^2 - 5$$

Now that we have the derivative, to find the slope of the original equation at $x = 0$, we just use 0 as our x value in the derivative:

$$y' = e^x + 3x^2 - 5$$
$$= e^0 + 3(0)^2 - 5$$
$$= 1 + 0 - 5$$
$$= -4$$

Therefore, the slope of the original equation at $x = 0$ is -4.

20. **Question:** For the equation $y = x^2 + x - 6$ find the x value where the slope is $\frac{3}{2}$.

    Solution: The equation has a slope of $\frac{3}{2}$ at $x = \frac{1}{4}$.

    Explanation: Since we are looking for a slope, the first operation to do is the derivative:

    $$y = x^2 + x - 6$$
    $$y' = 2x + 1$$

    We are looking for the x location where the *slope* (which will be y' in this equation) is $\frac{3}{2}$. Therefore, we simply replace y' with $\frac{3}{2}$ and solve for x:

    $$y' = 2x + 1$$
    $$\frac{3}{2} = 2x + 1$$
    $$\frac{1}{2} = 2x$$
    $$\frac{1}{4} = x$$

    Therefore, the equation $y = x^2 + x - 6$ has a slope of $\frac{3}{2}$ at $x = \frac{1}{4}$.

21. **Question:** For the equation $y = x^2 - 3x + 5$, find the equation of the line that is tangent to this graph (i.e., has the same slope as the graph and touches it at a single point) at $x = 4$.

    Solution: $y = 5x - 11$

    Explanation: This problem sounds hard, but it is actuall just combining ideas that you already know. A line can be known from either

two points, or from the slope and a single point. We can find the point from plugging $x = 4$ into the equation and finding the y value. Then, we can use the derivative to find the slope at this point. We can then use those two facts to find the equation of the line.

So, first, let's find the value of y at $x = 4$:

$$y = x^2 - 3x + 5$$
$$y = (4)^2 - 3(4) + 5$$
$$y = 16 - 12 + 5$$
$$y = 9$$

Therefore, the point will be $(4, 9)$. What is the slope at this point? To find the slope, we need to take the derivative:

$$y = x^2 - 3x + 5$$
$$y' = 2x - 3$$

The slope at $x = 4$ is therefore:

$$y' = 2x - 3$$
$$y' = 2(4) - 3$$
$$y' = 8 - 3$$
$$y' = 5$$

So, the slope is 5 and it passes through the point $(4, 9)$. So far, we know that the equation for the line is $y = 5x + b$. To solve for b, we just need to put in our values for x and y:

$$y = 5x + b$$
$$9 = 5(4) + b$$
$$9 = 20 + b$$
$$-11 = b$$

Therefore, the full equation is:

$$y = 5x - 11$$

This line is tangent to our original equation at $x = 4$.

22. **Question:** For the equation $y = e^x + x$ find the equation of the line tangent to this graph at

$x = 0$.

Solution: $y = 2x + 1$

Explanation: The equation of a line is $y = mx + b$. To find the equation of our line, we need the slope and a point on the graph. First, we will find the point on the graph by substituting $x = 0$ in the original equation:

$$y = e^x + x$$
$$y = e^0 + (0)$$
$$y = 1 + 0$$
$$y = 1$$

Therefore, the point of tangency is $(0, 1)$. Now we just need the slope. For that, we will take the derivative:

$$y = e^x + x$$
$$y' = e^x + 1$$

Now, we just substitute in 0 for x in the derivative to find the slope:

$$y' = e^x + 1$$
$$y' = e^0 + 1$$
$$y' = 2$$

So the derivative is 2. So far, we know the equation for the line is $y = 2x + b$. To find the value of b we simply substitute the known point for x and y:

$$y = 2x + b$$
$$1 = 2(0) + b$$
$$1 = b$$

Therefore, the equation for the line is $y = 2x + 1$.

Chapter 9

Basic Uses of the Derivative

To solve some of these problems you may need to refer to some of the standard geometry formulas in Appendix H.3.

1. **Question:** Find the point on the graph of $y = 3x^2 + 10x$ where the slope is 40.

   Solution: The slope is 40 at the point $(5, 125)$.

   Explanation: To find an equation for the slope, we first take the derivative:

   $$y = 3x^2 + 10x$$
   $$y' = 6x + 10$$

   Since this is the equation for the slope, we are just looking for where $y' = 40$.

   $$y' = 6x + 10$$
   $$40 = 6x + 10$$
   $$30 = 6x$$
   $$5 = x$$

   Therefore, the slope is 40 where $x = 5$. The question asked for the *point*, which implies both an x and a y value. Therefore, substituting 5 back into the equation we get:

   $$y = 3x^2 + 10x$$
   $$y = 3(5)^2 + 10(5)$$
   $$y = 3 \cdot 25 + 50$$
   $$y = 125$$

   Therefore, the slope is 40 at the point $(5, 125)$.

2. **Question:** What is the slope of the equation $y = 5x^4 + 3x - 5$ at $x = 2$?

   Solution: The slope at $x = 2$ is 163 (or, you could say, $\frac{163}{1}$).

   Explanation: To find the slope at a particular point, all we need to do is to take the derivative:

   $$y = 5x^4 + 3x - 5$$
   $$y' = 20x^3 + 3$$

   Now we substitute with $x = 2$:

   $$y' = 20x^3 + 3$$
   $$y' = 20(2)^3 + 3$$
   $$y' = 20 \cdot 8 + 3$$
   $$y' = 163$$

3. **Question:** Find the second derivative of $y = e^x + 3x^2$ at $x = 0$.

   Solution: The second derivative at $x = 0$ is 7.

   Explanation: The first derivative is:

   $$y = e^x + 3x^2$$
   $$y' = e^x + 6x$$

   The second derivative is:

   $$y' = e^x + 6x$$
   $$y'' = e^x + 6$$

At $x = 0$, this becomes:

$$y'' = e^x + 6$$
$$y'' = e^0 + 6$$
$$y'' = 1 + 6$$
$$y'' = 7$$

4. **Question:** Find the third derivative of $y = x^5 - x^3$.

   Solution: $y''' = 60x^2 - 6$

   Explanation: The third derivative can be found by simply taking the derivative three times:

   $$y = x^5 - x^3$$
   $$y' = 5x^4 - 3x^2$$
   $$y'' = 20x^3 - 6x$$
   $$y''' = 60x^2 - 6$$

For the equations below, find both the x and y values for all local maxima and minima *using the derivative* (don't just use the vertex formula even if it applies). For each point found, use the second derivative test to find out if the point is a local maximum or minimum.

5. **Question:** $y = x^2 + 3x + 10$

   Solution: The equation has a local minima at $x = \frac{3}{2}$, $y = \frac{67}{4}$).

   Explanation: To find the x value(s) of the local minima and maxima of this function, we need to take the derivative and then find the point where the derivative is equal to zero.

   $$y = x^2 + 3x + 10$$
   $$y' = 2x + 3 \qquad \text{power and addition rules}$$
   $$0 = 2x + 3 \qquad \text{find where } y' \text{ is zero}$$
   $$2x = -3 \qquad \text{solve for } x$$
   $$x = \frac{-3}{2}$$

So, the derivative is zero (maxima or minima) when x is $\frac{-3}{2}$. Now find the y value for this x using the original equation:

$$y = x^2 + 3x + 10$$
$$= (\frac{-3}{2})^2 + 3 \cdot \frac{-3}{2} + 10$$
$$= \frac{9}{4} - \frac{9}{2} + 10$$
$$= \frac{9}{4} - \frac{18}{4} + \frac{40}{4}$$
$$= \frac{31}{4}$$

Therefore, the maxima or minima occurs at $(\frac{-3}{2}, \frac{31}{4})$.

Is the value a minima or maxima? The second derivative test will tell us:

$$y' = 2x + 3 \qquad \text{first derivative}$$
$$y'' = 2 \qquad \text{second derivative}$$

The second derivative is positive, so the value is a minima.

6. **Question:** $y = x^3 + 7x^2 - 34x - 40$

   Solution: $(1.763, -72.705)$ is a local minima and $(-6.429, -376.462)$ is a local maxima (both points are decimal approximations).

   Explanation: To find the x value(s) of the local minima and maxima of this function, we need to take the derivative and then find the point where the derivative is equal to zero.

   $$y = x^3 + 7x^2 - 34x - 40$$
   $$y' = 3x^2 + 14x - 34 \qquad \text{first derivative}$$
   $$0 = 3x^2 + 14x - 34 \qquad \text{find where } y' \text{ is zero}$$
   $$x = \frac{-14 \pm \sqrt{14^2 - 4 \cdot 3 \cdot (-34)}}{2 \cdot 3} \qquad \text{quadratic formula}$$
   $$= \frac{-14 \pm \sqrt{196 + 408}}{6}$$
   $$= \frac{-14 \pm \sqrt{604}}{6}$$
   $$\approx \frac{-14 \pm 24.576}{6}$$
   $$\approx 1.763 \text{ or } -6.429$$

Therefore, this equation has a local maxima or minima at two x values. To find the y values, we substitute each one into the original equation:

$$y = x^3 + 7x^2 - 34x - 40$$
$$\approx (1.763)^3 + 7(1.763)^2 - 34(1.763) - 40$$
$$\approx 5.480 + 21.757 - 59.942 - 40$$
$$\approx -72.705$$

$$y = x^3 + 7x^2 - 34x - 40$$
$$\approx (-6.429)^3 + 7(-6.429)^2 - 34(-6.429) - 40$$
$$\approx -265.724 + -289.324 + 218.586 - 40$$
$$\approx -376.462$$

Therefore, our two points are $(1.763, -72.705)$ and $(-6.429, -376.462)$.

Now, are these maxima or minima? The second derivative test will tell us.

$$y = x^3 + 7x^2 - 34x - 40$$

$y' = 3x^2 + 14x - 34$	first derivative
$y'' = 6x + 14$	second derivative

$y'' \approx 6(1.763) + 14$	first x value
≈ 24.578	local minima

$y'' = 6x + 14$	second derivative
$\approx 6(-6.429) + 14$	second x value
≈ -24.574	local maxima

Therefore, $(1.763, -72.705)$ is a local minima and $(-6.429, -376.462)$ is a local maxima.

7. **Question:** $y = -6x^2 + 7x + 5$

   Solution: The point $(\frac{7}{12}, \frac{1014}{144})$ is a local maxima.

   Explanation: To find the x value(s) of the local minima and maxima of this function, we need to take the derivative and then find the point where the derivative is equal to zero.

$$y = -6x^2 + 7x + 5$$

$y' = -12x + 7$	first derivative
$0 = -12x + 7$	find where y' is zero
$12x = 7$	solve for x
$x = \dfrac{7}{12}$	

Therefore, this equation has a local maxima or minima at $x = \frac{7}{12}$. To find the y value, we substitute this value back into the original equation:

$$y = -6x^2 + 7x + 5$$
$$= -6(\frac{7}{12})^2 + 7(\frac{7}{12}) + 5$$
$$= \frac{-294}{144} + \frac{49}{12} + 5$$
$$= \frac{-294}{144} + \frac{588}{144} + \frac{720}{144}$$
$$= \frac{1014}{144}$$

Therefore, the point is at $(\frac{7}{12}, \frac{1014}{144})$. Is that a maxima or minima? We can use the second derivative test to find out.

$$y = -6x^2 + 7x + 5$$

$y' = -12x + 7$	first derivative
$y'' = -12$	second derivative

Since the second derivative is negative at this point (because it is negative at all points), that means that this is a local maxima.

8. **Question:** $y = 2x^2 + 8x + 6$

   Solution: A local minima occurs at $(-2, -2)$

   Explanation: To find the x value(s) of the local minima and maxima of this function, we need to take the derivative and then find the point where the derivative is equal to zero.

$$y = 2x^2 + 8x + 6$$

$y' = 4x + 8$	first derivative
$0 = 4x + 8$	find where y' is zero
$-4x = 8$	solve for x
$x = -2$	

Therefore, the derivative is zero (has a maxima or minima) when $x = -2$. Now find the y value for this x using the original equation:

$$y = 2x^2 + 8x + 6$$
$$= 2(-2)^2 + 8(-2) + 6$$
$$= 8 - 16 + 6$$
$$= -2$$

Therefore, the maxima or minima occurs at $(-2, -2)$.

The second derivative will tell if the value is a maxima or minima:

$$y = 2x^2 + 8x + 6$$
$$y' = 4x + 8 \qquad \text{first derivative}$$
$$y'' = 4 \qquad \text{second derivative}$$

Since the second derivative is positive at all points, the second derivative is positive at the x value in question, so the value is a minima.

9. **Question:** $y = 18x^3 + 102x^2 - 40x - 24$

   Solution: $(-3.9646, 616.1411)$ is a local maxima and $(0.1868, -27.7955)$ is a local minima.

   Explanation: To find the x value(s) of the local minima and maxima of this function, we need to take the derivative and then find the point where the derivative is equal to zero.

$$y = 18x^3 + 102x^2 - 40x - 24$$
$$y' = 54x^2 + 204x - 40 \qquad \text{first derivative}$$
$$0 = 54x^2 + 204x - 40 \qquad \text{find where } y' \text{ is zero}$$
$$x = \frac{-204 \pm \sqrt{204^2 - 4 \cdot 54 \cdot (-40)}}{2 \cdot 54} \qquad \text{quadratic formula}$$
$$= \frac{-204 \pm \sqrt{41616 + 8640}}{108}$$
$$\approx \frac{-204 \pm 224.1785}{108}$$
$$\approx -3.9646 \text{ or } 0.1868$$

Therefore, this equation has a local maxima or minima at two x values. To find the y values, we substitute each one into the original equation:

$$y = 18x^3 + 102x^2 - 40x - 24$$
$$= 18(-3.9646)^3 + 102(-3.9646)^2 - 40(-3.9646) - 24$$
$$\approx -1121.6843 + 1603.2414 + 158.584 - 24$$
$$\approx 616.1411$$

$$y = 18x^3 + 102x^2 - 40x - 24$$
$$= 18(0.1868)^3 + 102(0.1868)^2 - 40(0.1868) - 24$$
$$\approx 0.1173 + 3.5592 - 7.472 - 24$$
$$\approx -27.7955$$

Therefore, our two points are $(-3.9646, 616.1411)$ and $(0.1868, -27.7955)$

Now, are these maxima or minima? The second derivative test will tell us.

$$y = 18x^3 + 102x^2 - 40x - 24$$
$$y' = 54x^2 + 204x - 40 \qquad \text{first derivative}$$
$$y'' = 108x + 204 \qquad \text{second derivative}$$

$$y'' \approx 108(-3.9646) + 204 \qquad \text{first } x \text{ value}$$
$$\approx -224.1768 \qquad \text{local maxima}$$

$$y'' \approx 108(0.1868) + 204 \qquad \text{second } x \text{ value}$$
$$\approx 224.1744 \qquad \text{local minima}$$

Therefore $(-3.9646, 616.1411)$ is a local maxima and $(0.1868, -27.7955)$ is a local minima.

Solve the following problems *using derivatives*:

10. **Question:**   Find the global minimum and maximum of the equation $y = x^2$.

    Solution: The global minimum occurs at $(0, 0)$ and the global maximum is infinity and occurs at both ends of the graph at $(-\infty, +\infty)$ and $(+\infty, +\infty)$.

    Explanation: To find the global minimum and maximum, first find all of the critical points. The first critical points we will look for is the local minima and maxima of the equation using the derivative:

$$y = x^2$$

$$y' = 2x \qquad \text{take the derivative}$$

$$0 = 2x \qquad \text{find where it is zero}$$

$$x = 0$$

So, for $x = 0$, the y value is $0^2 = 0$. Since no limits were given for this equation, and it is continuous all the way through, we will think about where the function goes at the extremity of $x = \pm\infty$.

If you think about the graph, what shape is it? $y = x^2$ forms a parabola pointing up. Therefore, on each side, the values get bigger in an unbounded way as x continues towards $+\infty$ on the right, and $-\infty$ on the left. Since it is increasing unbounded, we say that the value for these is $+\infty$.

Therefore, our critical values are: $(0,0)$, $(+\infty, +\infty)$, and $(-\infty, +\infty)$. Since $(0,0)$ is the smallest value, it is our global minimum. Since the other two are both $+\infty$, they are both the global maximum.

11. **Question:** Take the equation $y = x^3 + 4x^2 + x - 6$ on the interval from $x = -4$ to $x = 2$. What is the maximum value on this interval?

Solution: The maximum value on this interval is 20, which occurs at the point $(2, 20)$.

Explanation: To find the maximum value, we need to find all of the critical values. To start with, we can find the local minima and maxima

of the function using the derivative.

$$y = x^3 + 4x^2 + x - 6$$

$$y' = 3x^2 + 8x + 1 \qquad \text{derivative}$$

$$0 = 3x^2 + 8x + 1 \qquad \text{find where derivative is zero}$$

$$x = \frac{-8 \pm \sqrt{(8)^2 - 4 \cdot 3 \cdot 1}}{2 \cdot 3} \qquad \text{solve}$$

$$= \frac{-8 \pm \sqrt{64 - 12}}{6}$$

$$= \frac{-8 \pm \sqrt{52}}{6}$$

$$\approx -0.1315 \text{ or } -2.5352$$

We now need to find the y values for these points:

$$y = x^3 + 4x^2 + x - 6$$

$$= (-0.1315)^3 + 4(-0.1315)^2 + (-0.1315) - 6$$

$$= -6.0646$$

$$y = x^3 + 4x^2 + x - 6$$

$$= (-2.5352)^3 + 4(-2.5352)^2 + (-2.5352) - 6$$

$$= 0.8794$$

Now we have two other points we need to look at—the beginning and the end of the graph. The problem defines these points as $x = -4$ and $x = 2$. So we will find the y values for these:

$$y = (-4)^3 + 4(-4)^2 + (-4) - 6$$

$$= -10$$

$$y = (2)^3 + 4(2)^2 + (2) - 6$$

$$= 20$$

So, our collection of critical points is $(-0.1315, -6.0646)$, $(-2.5352, 0.8794)$, $(-4, -10)$, and $(2, 20)$. We are looking for the maximum value, which will be in this list. Therefore, the maximum value occurs at the end of the interval, at $(2, 10)$.

12. **Question:** I am building a rectangular fence bordering a straight river (so I don't need fence on one side) and I have 200ft of fencing. First, establish an equation that finds the total

area enclosed based on one variable (probably one of the sides of the fence). Second, solve that equation to find out what the lengths of the sides need to be in order to maximize the amount of area enclosed by that fence.

Solution: The equation for the total area enclosed based on one variable is $A = 200L - 2L^2$. The fence can enclose a maximum of 5000 square feet using 200 feet of fencing with a length of 50 feet and a width of 100 feet.

Explanation: Since it is a rectangular fence, and we are talking about fencing it in, we are talking about the perimeter of the fence. The perimeter of a rectangle is $2L + 2W$ (where L is the size of the length and W is the size of the width). However, in this case, one side is missing (we will just say arbitrarily that it is the width), so it will be $2L + W$. The total perimeter is given by the amount of fencing that we have (200 feet). Therefore, we can establish the following equation relating length to height:

$$2L + W = 200$$

Now, the area of a rectangle (A) is always given by $A = L \cdot W$. This is an equation of *two* variables. To find the maxima, we need to take a derivative, but we need it as an equation of *one* variable to perform the derivative. Therefore, we will use the equation about the perimeter to reduce our variable count:

$$2L + W = 200W \qquad = 200 - 2L$$

Now that we have an equation for W in terms of L, we can substitute it in our equation for area:

$$A = L \cdot W$$
$$A = L \cdot (200 - 2L) \qquad \text{substituting for } W$$
$$A = 200L - 2L^2 \qquad \text{simplifying}$$

Now we have an equation with one independent variable (L) and one dependent variable (A), so we can use the derivative to find a local maxima/minima:

$$A = 200L - 2L^2$$
$$A' = 200 - 4L \qquad \text{derivative}$$
$$0 = 200 - 4L \qquad \text{find where derivative is zero}$$
$$4L = 200 \qquad \text{solve for } L$$
$$L = 50$$

Therefore, we have a local maxima or minima where $L = 50$. But which is it? The second derivative test will tell us:

$$A = 200L - 2L^2$$
$$A' = 200 - 4L \qquad \text{first derivative}$$
$$A'' = -4 \qquad \text{second derivative}$$

Therefore, since the second derivative is negative, this is indeed a maxima.

Since this is a downward-pointing parabola (since it is a quadratic), we know that this local maxima is actually the global maximum. However, to follow the full process, we should look at the other critical points involved.

The equation has two other critical points. Since L is a length, L cannot go below zero. Additionally, because W (i.e., $200 - 2L$) is a length, then $200 - 2L$ can't go below zero, either (this happens when $L = 100$). The area at these two locations is zero, so these are obviously not eligible to be the maximum. Because the equation is bounded between $L = 0$ and $L = 100$, we don't need to examine what the equation is doing when L is at positive or negative infinity.

Therefore, this leaves $L = 50$ as the only possibility for a global maximum.

Now, we can find the other variables:

$$W = 200 - 2L$$
$$= 200 - 2(50)$$
$$= 200 - 100$$
$$= 100$$

$$A = L \cdot W$$
$$= 100 \cdot 50$$
$$= 5000$$

Therefore, the fence can enclose a maximum of 5000 square feet using 200 feet of fencing with a length of 50 feet and a width of 100 feet.

13. **Question:** Imagine a cardboard box with a **square** base that has no top. Find the dimensions of the box with the largest volume that can be made out of 35 square feet of cardboard.

Solution: The dimensions of the box are $L \approx 3.4157$ feet, $W \approx 3.4157$ feet, and $H \approx 1.7078$ feet, giving a total volume of 19.9246 cubic feet.

Explanation: The volume of a box can be given by the following formula, where V is volume, L is length, W is width, and H is height:

$$V = L \cdot W \cdot H$$

Now, the total amount of cardboard tells us about the *surface area* of the box, not about its volume. Surface area is calculated by adding up together each side of the box. The equation for surface area is below (we only have one LW because the top is open):

$$A = LW + 2WH + 2LH$$

Now, the question specifies a *square* base. This means that two of our dimensions will be the same. Therefore, we will simply say that $W = L$. Therefore, in each formula, we can replace W with L. This gives us:

$$V = L \cdot W \cdot H$$
$$= L^2 \cdot H$$

$$A = LW + 2WH + 2LH$$
$$= LL + 2LH + 2LH$$
$$= L^2 + 4LH$$

Since we are trying to maximize the volume, we need to take the derivative of it. However, we have *two* independent variables (L and H) and we need to reduce it. However, since we know the surface area (i.e., the total amount of cardboard to use—35), we can use the surface area equation to solve for one of the variables. H will be the easiest one to solve for:

$$A = L^2 + 4LH$$
$$35 = L^2 + 4LH$$
$$35 - L^2 = 4LH$$
$$H = \frac{35 - L^2}{4L}$$

Now, we can substitute this in to the volume equation as follows:

$$V = L^2 \cdot H$$
$$= L^2 \left(\frac{35 - L^2}{4L} \right)$$
$$= \frac{L^2 (35 - L^2)}{4L}$$
$$= \frac{35L^2 - L^4}{4L}$$
$$V = \frac{35}{4} L - \frac{1}{4} L^3$$

Now we have an equation in a single variable as a polynomial. Therefore, we can find any local maxima and minima by finding where the derivative is zero:

$$V = \frac{35}{4} L - \frac{1}{4} L^3$$

$$V' = \frac{35}{4} - \frac{3}{4} L^2 \qquad \text{first derivative}$$

$$0 = \frac{35}{4} - \frac{3}{4} L^2 \qquad \text{find where derivative is zero}$$

$$0 = -\frac{3}{4} L^2 + 0L + \frac{35}{4} \qquad \text{form for quadratic formula}$$

$$L = \frac{0 \pm \sqrt{0 - 4 \cdot \frac{-3}{4} \cdot \frac{35}{4}}}{2 \cdot \frac{-3}{4}} \qquad \text{quadratic formual}$$

$$= \pm \frac{\sqrt{105}}{3}$$

$$\approx \pm 3.4157$$

So, as we can see, the extrema of the volume will be where the length is either positive or negative 3.4157. Obviously, the length cannot be a negative number. That leaves us with a length of 3.4157.

Is that a maxima or a minima? The second derivative test will tell us:

$$V' = \frac{35}{4} - \frac{3}{2} L^2 \qquad \text{first derivative}$$

$$V'' = -3L \qquad \text{second derivative}$$

Since L is positive, $-3L$ will be negative, and that means that it is a maxima.

However, just because it is a *local* maxima does not mean that it is a *global* maxima. Is this a global maxima? We need to check the other critical points to be sure.

Let's think about how the graph will look. If it has a maxima on the right and a minima on the left, that means that to the right of the maxima it will trend downward (towards negative infinity), but to the left of the minima it will trend upward towards positive infinity. However, the minima occurs where L is negative, and L cannot validly be a negative number. Therefore, we can ignore the trend towards positive infinity on the left, because it doesn't start until L is an invalid value.

On the right-hand side, we can calculate the maximum valid value before another dimension turns negative, but we don't need to, since the graph is sloping downwards. Therefore, we can disregard the left side of the graph's uptrend (because it happens where L is negative) and we can disregard the graph to the right of the maximum (because it is on a downtrend), and are left with our local maxima as the global maximum.

So, at $L = 3.4157$ we have the maximum possible volume for the box. Using the equation for the height we find that:

$$H = \frac{35 - L^2}{4L}$$
$$\approx \frac{35 - (3.4157)^2}{4 \cdot 3.4157}$$
$$\approx 1.7078$$

The width is the same as the height, so $W = 3.4157$. The maximum volume itself can be found using our volume equation:

$$V = \frac{35}{4}L - \frac{1}{4}L^3$$
$$\approx \frac{35}{4}3.4157 - \frac{1}{4}(3.4157)^3$$
$$\approx 19.9246$$

14. **Question:** Create an equation for the previous question so that it can be solved for any given amount of cardboard using just a calculator.

Solution: If A is the total area of cardboard

allowed to be used, then:

$$L = \frac{\sqrt{3A}}{3}$$
$$W = \frac{\sqrt{3A}}{3}$$
$$H = \frac{A}{2\sqrt{3A}}$$
$$V = \frac{A^2}{6\sqrt{3A}}$$

Note that these are simplified versions. It is not an error to not have simplified it all the way. You can check your own answers by substituting 35 in for A and seeing if you get the same results as in the previous question.

Explanation: The way that we can generalize an equation is by substituting a given value with a constant-variable. Therefore, we will use A (the surface area) as the constant variable. Remember that when you do this, A gets treated as a *constant* for the purposes of the derivative.

Therefore, we have the equations:

$$V = L^2 H$$
$$A = L^2 + 4LH$$

We can then solve for H as follows:

$$A = L^2 + 4LH$$
$$4LH = A - L^2$$
$$H = \frac{A - L^2}{4L}$$

Therefore, the volume equation becomes:

$$V = L^2 \frac{A - L^2}{4L}$$
$$V = \frac{AL^2 - L^4}{4L}$$
$$V = \frac{A}{4}L - \frac{1}{4}L^3$$

Now we can take the derivative and find where

it is zero:

$$V = \frac{A}{4}L - \frac{1}{4}L^3$$

$$V' = \frac{A}{4} - \frac{3}{4}L^2$$

$$0 = -\frac{3}{4}L^2 + 0L + \frac{A}{4}$$

$$L = \frac{0 \pm \sqrt{0 - 4 \cdot (\frac{-3}{4}) \cdot \frac{A}{4}}}{2 \cdot \frac{-3}{4}}$$

$$= \pm \frac{2\sqrt{\frac{3A}{4}}}{3}$$

$$= \pm \frac{\sqrt{3A}}{3}$$

$$= \frac{\sqrt{3A}}{3}$$

In the last line of the above, we removed the ± because the length can *only* be positive, and we showed in the previous problem why this is a global maximum.

Now, if the length is given by $\frac{\sqrt{3A}}{3}$, then that is also the width. The height can be found by:

$$H = \frac{A - L^2}{4L}$$

$$= \frac{A - (\frac{\sqrt{3A}}{3})^2}{4 \cdot \frac{\sqrt{3A}}{3}}$$

$$= \frac{A - \frac{3A}{9}}{4 \cdot \frac{\sqrt{3A}}{3}}$$

$$= \frac{\frac{9A-3A}{9}}{4 \cdot \frac{\sqrt{3A}}{3}}$$

$$= \frac{3}{4\sqrt{3A}} \cdot \frac{6A}{9}$$

$$= \frac{A}{2\sqrt{3A}}$$

Finally, if we wanted to, we can determine the area of the final box as:

$$V = L^2 H$$

$$= \left(\frac{\sqrt{3A}}{3}\right)^2 \cdot \frac{A}{2\sqrt{3A}}$$

$$= \frac{3A}{9} \cdot \frac{A}{2\sqrt{3A}}$$

$$= \frac{A^2}{6\sqrt{3A}}$$

If you wanted to verify this, you can plug in 35 for A to see if the values match the previous problem.

15. **Question:** A drink company is wanting to minimize the amount of material needed to make a can for their product. A standard drink serving is 355 mL (1 mL is a cubic centimeter). What dimensions should they make the can (which is a cylinder) in order to minimize the material needed (i.e., the surface area)?

Solution: The can that will use the minimal surface area will have a radius of 3.8372 centimeters and a height of 7.6745 centimeters.

Explanation: The surface area of a cylinder is given by the following equation:

$$a = 2\pi r h + 2\pi r^2$$

However, this has *two* variables, not just one. To reduce the number of variables, we can use the other fact given—that the volume will be 355 cubic centimeters. The equation for the volume of a cylinder is:

$$v = \pi r^2 h$$

Since we know the volume (355), we can substitute that in and get:

$$355 = \pi r^2 h$$

Now, we can solve for h simply by dividing both sides by πr^2:

$$h = \frac{355}{\pi r^2}$$

We can subsitute this into the surface area

equation as follows:

$$a = 2\pi r h + 2\pi r^2$$

$$= 2\pi r \left(\frac{355}{\pi r^2}\right) + 2\pi r^2$$

$$= \frac{2 \cdot 355\pi r}{\pi r^2} + 2\pi r^2$$

$$= \frac{710}{r} + 2\pi r^2 \qquad = 710r^{-1} + 2\pi r^2$$

Local minimas and maximas occur when the derivative is zero. Therefore, we will start by taking the derivative:

$$a = 710r^{-1} + 2\pi r^2$$

$$a' = -710r^{-2} + 4\pi r$$

Next, set the derivative to zero, and solve for where the derivative is zero (we will multiply by r^2 in order to get rid of r in the denominator):

$$a' = -710r^{-2} + 4\pi r$$

$$0 = -710r^{-2} + 4\pi r$$

$$0 = -710 + 4\pi r^3$$

$$710 = 4\pi r^3$$

$$\frac{710}{4\pi} = r^3$$

$$\frac{355}{2\pi} = r^3$$

$$r = \sqrt[3]{\frac{355}{2\pi}}$$

$$r \approx 3.8372$$

We can verify that this is a local minima using the second derivative test. First, we will use the first derivative to find the second derivative:

$$a' = -710r^{-2} + 4\pi r$$

$$a'' = 1420r^{-3} + 4\pi$$

You can see from this that, for any positive r, the second derivative will be positive, so that means that the value is indeed a minima.

Now that we have the value for the radius that we want, we need to find the corresponding height. To find its height, we just use the equation above:

$$h = \frac{355}{\pi r^2}$$

$$= \frac{355}{\pi (3.8372)^2}$$

$$= 7.6745$$

Since the units of the original question were in centimeters, these values are also in centimeters. Therefore, the can that will use the minimal surface area will have a radius of 3.8372 centimeters and a height of 7.6745 centimeters.

16. **Question:** Find the closed cylinder (i.e., height and radius) with the largest volume that can be made out of 200 square feet of paper.

Solution: The largest closed cylinder that can be made with 200 square feet of paper has a radius of 3.2574 feet, a height of 6.5145 feet, and final volume of 217.1567 cubic feet.

Explanation: This is very similar to the previous problems, but with a different shape—a cylinder. So, to start off with, we are going to start with the equation for the volume of a cylinder:

$$V = \pi R^2 H$$

Again, this is an equation with two independent variables. So, we have to use the surface area information to reduce the number of variables.

$$A = 2\pi R H + 2\pi R^2$$

Because we know the surface area (200 square feet), we can solve for the height to reduce the number of variables:

$$200 = 2\pi R H + 2\pi R^2$$

$$2\pi R H = 200 - 2\pi R^2$$

$$H = \frac{200 - 2\pi R^2}{2\pi R}$$

$$H = \frac{100}{\pi R} - R$$

Therefore, we can substitute this back into the original volume equation:

$$V = \pi R^2 H$$

$$V = \pi R^2 \left(\frac{100}{\pi R} - R\right)$$

$$V = 100R - \pi R^3$$

We now have a polynomial with a single independent variable, and can therefore calculate a derivative with (remember that π is a constant):

$$V = 100R - \pi R^3$$

$$V' = 100 - 3\pi R^2$$

$$0 = 100 - 3\pi R^2 \qquad \text{find where slope is zero}$$

$$0 = -3\pi R^2 + 0R + 100 \qquad \text{put in quadratic form}$$

$$R = \frac{0 \pm \sqrt{0^2 - 4(-3\pi)(100)}}{2 \cdot -3\pi} \qquad \text{quadratic formula}$$

$$= \frac{\pm\sqrt{1200\pi}}{6\pi} \qquad \begin{array}{l}\text{note that the}\\\text{bottom negative}\\\text{is irrelevant}\\\text{because of } \pm \text{ on}\\\text{the top}\end{array}$$

$$= \frac{\pm 20\sqrt{3\pi}}{6\pi}$$

$$= \frac{\pm 10\sqrt{3\pi}}{3\pi}$$

$$= \frac{\pm 10\sqrt{3\pi}}{3\pi}$$

$$\approx \pm 3.2574$$

$$\approx 3.2574 \qquad \text{only positive lengths}$$

Therefore, we have an extremum at 3.2574. Is it a local maxima or a local minima? The second derivative will tell us:

$$V = 100R - \pi R^3$$

$$V' = 100 - 3\pi R^2$$

$$V'' = -6\pi^2 R$$

For any positive R, this will be negative, and therefore it will be a local maxima. Just like for the previous problems, this function *can* go infinitely high, but *only* for negative R values, which don't work in this function, since lengths cannot be negative.

Therefore, we can substitute this R value in and find the height and the final volume at the maximum:

$$V = 100R - \pi R^3$$

$$\approx 100(3.2574) - \pi(3.2574)^3$$

$$\approx 217.1567$$

$$H = \frac{100}{\pi R} - R$$

$$\approx \frac{100}{\pi(3.2574)} - (3.2574)$$

$$\approx 6.5145$$

17. **Question:** Create an equation for the previous question so that it can be solved for any given amount of paper using just a calculator.

Solution: The equations for maximum radius, height, and volume, given a surface area of A are:

$$R = \frac{\sqrt{6 \cdot A \cdot \pi}}{6\pi}$$

$$H = \frac{3A}{\sqrt{6A\pi}} - \frac{\sqrt{6A\pi}}{6\pi}$$

$$V = \frac{A}{2}\left(\frac{\sqrt{6A\pi}}{6\pi}\right) - \pi\left(\frac{\sqrt{6A\pi}}{6\pi}\right)^3$$

Explanation: To solve this, we will use A to represent the total area of paper allowed to be used. Therefore, we can say:

$$A = 2\pi RH + 2\pi R^2$$

$$2\pi RH = A - 2\pi R^2$$

$$H = \frac{A - 2\pi R^2}{2\pi R}$$

$$= \frac{A}{2\pi R} - R$$

Now we can formulate the volume in terms of R

only:

$$V = \pi R^2 H$$

$$V = \pi R^2 \left(\frac{A}{2\pi R} - R \right)$$

$$V = \frac{A\pi R^2}{2\pi R} - \pi R^3$$

$$V = \frac{A}{2}R - \pi R^3$$

Now we can take the derivative and find where the derivative is zero (remember that both π and A are constants in this):

$$V = \frac{A}{2}R - \pi R^3$$

$$V' = \frac{A}{2} - 3\pi R^2$$

$$0 = \frac{A}{2} - 3\pi R^2$$

$$0 = -3\pi R^2 + 0R + \frac{A}{2} \qquad \text{quadratic form}$$

$$R = \frac{0 \pm \sqrt{0^2 - 4 \cdot (-3\pi) \cdot \frac{A}{2}}}{2 \cdot -3\pi}$$

$$= \frac{\sqrt{6A\pi}}{6\pi}$$

The height will be:

$$H = \frac{A}{2\pi \left(\frac{\sqrt{6A\pi}}{6\pi} \right)} - \frac{\sqrt{6A\pi}}{6\pi}$$

$$= \frac{A}{\left(\frac{\sqrt{6A\pi}}{3} \right)} - \frac{\sqrt{6A\pi}}{6\pi}$$

$$= \frac{3A}{\sqrt{6A\pi}} - \frac{\sqrt{6A\pi}}{6\pi}$$

The final volume is therefore:

$$V = \frac{A}{2} \left(\frac{\sqrt{6A\pi}}{6\pi} \right) - \pi \left(\frac{\sqrt{6A\pi}}{6\pi} \right)^3$$

18. **Question:** Just like there is a vertex formula for quadratics, develop a vertex formula for cubic

equations (i.e., equations where the highest power of x is 3).

Solution: $\frac{-B \pm \sqrt{B^2 - 3AC}}{3A}$

Explanation: When we derived the vertex formula for quadratics, we simply solved the problem using A, B, and C as constant variables. Here we will do the same, but we will also add in D as well. The basic form of the equation is as follows:

$$y = Ax^3 + Bx^2 + Cx + D$$

The vertices, by definition, are at local minima and maxima. Therefore, to find the minima and maxima, we take the derivative and find where it is zero (don't forget that the constant-variables act as *constants*!):

$$y = Ax^3 + Bx^2 + Cx + D$$

$$y' = 3Ax^2 + 2Bx + C \qquad \text{derivative}$$

$$0 = 3Ax^2 + 2Bx + C \qquad \begin{matrix}\text{find where} \\ \text{derivative is} \\ \text{zero}\end{matrix}$$

$$x = \frac{-(2B) \pm \sqrt{(2B)^2 - 4 \cdot 3A \cdot C}}{2 \cdot 3A} \qquad \begin{matrix}\text{quadratic} \\ \text{formula}\end{matrix}$$

$$= \frac{-2B \pm \sqrt{4B^2 - 12AC}}{6A} \qquad \text{simplifying}$$

$$= \frac{-2B \pm \sqrt{4(B^2 - 3AC)}}{6A}$$

$$= \frac{-2B \pm 2\sqrt{B^2 - 3AC}}{6A}$$

$$= \frac{-2B \pm 2\sqrt{B^2 - 3AC}}{6A}$$

$$= \frac{-B \pm \sqrt{B^2 - 3AC}}{3A}$$

Since we are looking for a formula to find just the x values, this works as the formula.

As a side note, if both values for x are the same, this will actually not be a minima or a maxima. Also note that developing vertex equations for higher-degreed polynomials is exceedingly more difficult.

19. **Question:**   Use the formula derived in the previous question to find the x values of the vertices of the following equation: $y = x^3 + 5x^2 + 3x + 2$. Also find the corresponding

y values of the vertices.

Solution: The vertices are at $(-\frac{1}{3}, \frac{41}{27})$ and $(-3, 11)$.

Explanation: In the previous question, we defined the vertex formula as being:

$$x = \frac{-B \pm \sqrt{B^2 - 3AC}}{3A}$$

Therefore, substituting the values for our equation, we get:

$$
\begin{aligned}
x &= \frac{-B \pm \sqrt{B^2 - 3AC}}{3A} \\
&= \frac{-(5) \pm \sqrt{(5^2) - 3 \cdot 1 \cdot 3}}{3 \cdot 1} \\
&= \frac{-5 \pm \sqrt{25 - 9}}{3} \\
&= \frac{-5 \pm \sqrt{16}}{3} \\
&= \frac{-5 \pm 4}{3} \\
&= -\frac{1}{3} \text{ or } -3
\end{aligned}
$$

Now, to find the *y* values, we need to substitute these back in:

$$
\begin{aligned}
y &= x^3 + 5x^2 + 3x + 2 && \text{original equation} \\
&= (-\tfrac{1}{3})^3 + 5(-\tfrac{1}{3})^2 + 3(-\tfrac{1}{3}) + 2 && \text{Using } -\tfrac{1}{3} \\
&= \frac{-1}{27} + \frac{5}{9} - 1 + 2 && \text{simplifying} \\
&= \frac{-1}{27} + \frac{15}{27} + \frac{27}{27} \\
&= \frac{41}{27}
\end{aligned}
$$

And, using the other value:

$$
\begin{aligned}
y &= x^3 + 5x^2 + 3x + 2 && \text{original equation} \\
&= (-3)^3 + 5(-3)^2 + 3(-3) + 2 && \text{Using } -3 \\
&= -27 + 45 - 9 + 2 && \text{simplifying} \\
&= 11
\end{aligned}
$$

The vertices are at $(-\frac{1}{3}, \frac{41}{27})$ and $(-3, 11)$.

20. **Question:** A smart phone manufacturer decides that in order to sell *n* units of their new cell phone, the wholesale price (*w*) per unit must be $(300 - w) \cdot 1,000,000$. The cost to produce the unit is \$150 per unit. Profit (*p*) is the difference between revenue and cost. What price should they set their cell phone for maximum profit, and what profit will they get?

Solution: The manufacturer will maximize their profit by selling at \$225 per unit, and the profit will be \$5,625,000,000.

Explanation: To solve this, first we need to combine our different facts into a single equation. Which variable are we trying to optimize for? We are trying to optimize profit. Which thing are we trying to vary? We are trying to vary price.

So, let's start by arranging our knowledge into a set of equations. The number of units sold is represented by the following equation:

$$n = (300 - w) \cdot 1,000,000$$

The revenue (which we can represent with *r*) is simply the price *w* multiplied by the number of units sold:

$$r = nw$$

The total cost (which we can represent with *c*) is simply the units multiplied by the cost per unit (\$150):

$$c = 150n$$

Profit is simply revenue minus cost:

$$p = r - c$$

What we want to do is optimize the profit, so that will be our final dependent variable. The variable we can control directly (i.e., the independent variable) will be the wholesal price (*w*). Therefore, we want to perform substitutions until we can get an equation with two variables: *w* as the independent variable and *p* as the dependent variable.

We can easily combine the profit, revenue, and

cost formulas:

$$p = r - c$$

$$r = nw$$

$$c = 150n$$

$$p = nw - 150n$$

$$p = n(w - 150)$$

This leaves two independent variables. However, the number of units is a function of the wholesale cost. That allows us to do one more substitution:

$$p = n(w - 150)$$

$$n = (300 - w) \cdot 1,000,000$$

$$p = ((300 - w) \cdot 1,000,000)(w - 150)$$

This can be simplified into a basic polynomial:

$$p = ((300 - w) \cdot 1,000,000)(w - 150)$$

$$p = 1,000,000(300 - w)(w - 150)$$

$$p = 1,000,000(300w - 45,000 - w^2 + 150w)$$

$$p = 1,000,000(-w^2 + 450w - 45,000)$$

$$p = -1,000,000w^2 + 450,000,000w - 45,000,000,000$$

To optimize this problem, we simply take the derivative:

$$p = -1,000,000w^2 + 450,000,000w - 45,000,000,000$$

$$p' = -2,000,000w + 450,000,000$$

An extrema occurs when the derivative is zero:

$$p' = -2,000,000w + 450,000,000$$

$$0 = -2,000,000w + 450,000,000$$

$$2,000,000w = 450,000,000$$

$$w = 225$$

To find if this is a minima or maxima, we take the second derivative:

$$p' = -2,000,000w + 450,000,000$$

$$p'' = -2,000,000$$

Because the second derivative is negative, this is indeed a maximal If you graph the equation, you will find that the curve is simply an inverted parabola, so this is also a global maxima.

Finally, we need to substitute our wholesale price back into the equation to find the final profit:

$$p = -1,000,000w^2 + 450,000,000w - 45,000,000,000$$

$$p = -1,000,000(225^2) + 450,000,000(225) - 45,000,000,000$$

$$p = 5,625,000,000$$

The total profit will be $5,625,000,000.

Chapter 10

The Differential

1. **Question:** Write down the differential rules listed in Figure 10.2 three times.

    Solution: The rule list is:

    Constant Rule $\mathrm{d}(C) = 0$
    Constant Multiplier Rule $\mathrm{d}(nx) = n\,\mathrm{d}x$
    Differential of a variable $\mathrm{d}(x) = dx$
    Power Rule $\mathrm{d}(x^n) = nx^{n-1}\,\mathrm{d}x$
    Exponent Rule $\mathrm{d}(n^x) = \ln(n)n^x\,\mathrm{d}x$
    Addition Rule $\mathrm{d}(f(x)+g(x)+\ldots) = \mathrm{d}(f(x)) + \mathrm{d}(g(x)) + \ldots$
    Addition Rule 2 $\mathrm{d}(x + y + z + \ldots) = \mathrm{d}x + \mathrm{d}y + \mathrm{d}z + \ldots$

2. **Question:** Take the differential of $y = x^6$.

    Solution: $\mathrm{d}y = 6x^5\,\mathrm{d}x$

    Explanation: The differential is very similar to the derivative, except that it is applied to each side independently. Therefore, we can do:

    $$\mathrm{d}(y) = \mathrm{d}(x^6)$$

    On the left, we have one term to take the differential of, y. The differential of any variable *variablename* is simply d*variablename*. Therefore, the differential of y is $\mathrm{d}y$.

    On the right, we have the differential of x^6. Using the power rule, we see that $\mathrm{d}(x^n) = nx^{n-1}\,\mathrm{d}x$. Therefore, the differential is $6x^5\,\mathrm{d}x$. That means the full equation is:

    $$\mathrm{d}y = 6x^5\,\mathrm{d}x$$

3. **Question:** Take the differential of $y = e^x$.

    Solution: $\mathrm{d}y = e^x\,\mathrm{d}x$

    Explanation: This uses the exponent rule of the differential:

    $$\begin{aligned} y &= e^x && \text{original equation} \\ \mathrm{d}(y) &= \mathrm{d}(e^x) && \text{differentiate both sides} \\ \mathrm{d}y &= e^x\,\mathrm{d}x && \text{exponent rule} \end{aligned}$$

4. **Question:** Take the differential of $y = 4x^3$.

    Solution: $\mathrm{d}y = 12x^2\,\mathrm{d}x$

    Explanation: This is another instance of the power rule:

    $$\begin{aligned} y &= 4x^3 && \text{original equation} \\ \mathrm{d}(y) &= \mathrm{d}(4x^3) && \text{differentiate both sides} \\ \mathrm{d}y &= 12x^2\,\mathrm{d}x && \text{power rule} \end{aligned}$$

5. **Question:** Take the differential of $y = 2x^2 + 5x + 4$.

    Solution: $\mathrm{d}y = 4x\,\mathrm{d}x + 5\,\mathrm{d}x$

    Explanation: In this equation, we are going to use the addition rule. This means that, at the + signs, we can take the differential of each part separately.

    On the left-hand side, we just have y, so the differential is $\mathrm{d}y$.

On the right-hand side, we have $2x^2 + 5x + 4$. The differential of $2x^2$ is $4x\,dx$. The differential of $5x$ is $5\,dx$. The differential of 4 is 0.

Therefore, the final differential is $dy = 4x\,dx + 5\,dx$.

6. **Question:** Take the differential of $y = \frac{5}{\sqrt{x}}$.

   Solution: $dy = \frac{-5}{2}x^{\frac{-3}{2}}\,dx$

   Explanation: To find this differential, we have to first rewrite it in terms of powers of x, and then take the differential.

   $$y = \frac{5}{\sqrt{x}}$$
   $$y = 5x^{\frac{-1}{2}}$$
   $$d(y) = d(5x^{\frac{-1}{2}})$$
   $$dy = 5 \cdot \frac{-1}{2}x^{\frac{-3}{2}}\,dx \qquad dy = \frac{-5}{2}x^{\frac{-3}{2}}\,dx$$

7. **Question:** Take the differential of $y^3 = 2x^2$.

   Solution: $3y^2\,dy = 4x\,dx$

   Explanation: This uses the power rule *on both sides of the equation.*

   To start with, we will take the differential of both sides. Therefore, it will be worked like this:

   $$y^3 = 2x^2 \qquad \text{original equation}$$
   $$d(y^3) = d(2x^2) \qquad \text{differentiate both sides}$$
   $$3y^2\,dy = 4x\,dx \qquad \text{apply the power rule}$$

   Note that the y side and the x side are treated *identically.* Don't forget that since the expression that we are taking the differential of on the left side is in terms of y, it uses the differential dy and **not** dx.

8. **Question:** Take the differential of $y^2 - x^2 + e^x = 3$.

   Solution: $2y\,dy - 2x\,dx + e^x\,dx = 0$

Explanation: This might look funny, because x and y are on the same side of the equation. But, with differentials, we can deal with this just fine. On the left-hand side, we are just using the addition rule, so we will take the differential of each component separately:

$$y^2 - x^2 + e^x = 3 \qquad \text{original equation}$$
$$d(y^2 - x^2 + e^x) = d(3) \qquad \text{differentiate}$$
$$d(y^2) - d(x^2) + d(e^x) = d(3) \qquad \text{addition rule}$$

From here we can apply the differential separately:

- The differential of y^2 is $2y\,dy$.
- The differential of x^2 is $2x\,dx$.
- The differential of e^x is $e^x\,dx$.

Therefore, the differential of the whole left side is $2y\,dy - 2x\,dx + e^x\,dx$.

On the right-hand side, the differential of any constant is zero. Therefore, the final equation is:

$$2y\,dy - 2x\,dx + e^x\,dx = 0$$

9. **Question:** Find the derivative of $2y + 5 = 5x - 3$.

   Solution: $\frac{dy}{dx} = \frac{5}{2}$

   Explanation: To find the derivative, we can simply take the differential and solve for $\frac{dy}{dx}$. Therefore, we will start by taking the differential:

   $$2y + 5 = 5x - 3 \qquad \text{original equation}$$
   $$d(2y + 5) = d(5x - 3) \qquad \text{differentiate}$$
   $$d(2y) + d(5) = d(5x) - d(3) \qquad \text{addition rule}$$
   $$2\,dy + 0 = 5\,dx + 0$$
   $$2\,dy = 5\,dx$$

   Now, we need to solve for $\frac{dy}{dx}$:

   $$2\,dy = 5\,dx \qquad \text{original differential}$$
   $$dy = \frac{5\,dx}{2} \qquad \text{solve for } \frac{dy}{dx}$$
   $$\frac{dy}{dx} = \frac{5}{2}$$

And that is our derivative. Note that the value is just a number because the original equation is just a line.

10. **Question:** Find the derivative of $e^y - x^3 = 2x$.

Solution: $\frac{dy}{dx} = \frac{2+3x^2}{e^y}$

Explanation: To find the derivative, we need to take the differential, and then solve for $\frac{dy}{dx}$:

$$e^y - x^3 = 2x \qquad \text{original equation}$$

$$e^y\,dy - 3x^2\,dx = 2\,dx \qquad \text{differential}$$

$$e^y\,dy = 2\,dx + 3x^2\,dx \qquad \text{separate } dy \text{ and } dx$$

$$e^y\,dy = (2 + 3x^2)\,dx \qquad \text{factor out } dx$$

$$dy = \frac{(2 + 3x^2)\,dx}{e^y} \qquad \text{solve for } \frac{dy}{dx}$$

$$\frac{dy}{dx} = \frac{2 + 3x^2}{e^y}$$

11. **Question:** Find the derivative of $3y^3 - 2x^2 + 5y + 20 = e^x$.

Solution: $\frac{dy}{dx} = \frac{e^x+4x}{9y^2+5}$

Explanation: To find the derivative, first we need to find the differential and then solve for $\frac{dy}{dx}$. The differential of this equation is:

$$9y^2\,dy - 4x\,dx + 5\,dy = e^x\,dx$$

Next, we need to get all of the dx terms on one side and the dy terms on the other:

$$9y^2\,dy + 5\,dy = e^x\,dx + 4x\,dx$$

Now, we can factor out the dy and dx from the rest of the terms:

$$(9y^2 + 5)\,dy = (e^x + 4x)\,dx$$

Now, cross-division will give us the solution:

$$\frac{dy}{dx} = \frac{e^x + 4x}{9y^2 + 5}$$

12. **Question:** The equation $e^y - 3x^2 = 5$ has one non-infinite minimum value for y that occurs at a local minima (you may use this as an assumption—if you find a local extremum, you can assume it is the value you are looking for). At what value of x does this occur? What is the y value at this point?

Solution: The minimum occurs at $x = 0$. The y value at this point is 1.6094.

Explanation: To get the minimum value, you need to find where the derivative is zero. The derivative is $\frac{dy}{dx}$, but, in this equation, since x and y are on the same side, it takes a little more effort.

To start, take the differential of the equation. That gives:

$$e^y\,dy - 6x\,dx = 0$$

Now, we need to solve for the derivative, which is $\frac{dy}{dx}$. In other words, one side of the equation needs to be exactly $\frac{dy}{dx}$ (and the other side needs to not include dy or dx). So, first we can split the x and y variables by adding $6x\,dx$ to both sides:

$$e^y\,dy = 6x\,dx$$

Next, we can divide both sides by e^y to get dy all by itself:

$$dy = \frac{6x\,dx}{e^y}$$

Now, we can divide both sides by dx to get $\frac{dy}{dx}$ on a side by itself:

$$\frac{dy}{dx} = \frac{6x}{e^y}$$

The minimum value occurs where the derivative, $\frac{dy}{dx}$, is zero. Therefore, we can simply set that side of the equation to zero:

$$0 = \frac{6x}{e^y}$$

We can multiply both sides by e^y to simplify:

$$0 = 6x$$

Now we can solve directly by dividing both sides by 6:

$$x = 0$$

Therefore, the minimum value occurs at $x = 0$. To get the y value, we just need to plug in x into the original equation, and then solve for y:

$$e^y - 3x^2 = 5$$
$$e^y - 3(0)^2 = 5$$
$$e^y - 0 = 5$$
$$e^y = 5$$
$$\ln(e^y) = \ln(5)$$
$$y = \ln(5)$$
$$y \approx 1.6094$$

13. **Question:**   Find the slope of the equation $y^2 = 4x^3$ at the point $(4, 16)$

    Solution: The slope of the graph at $(4, 16)$ is 6.

    Explanation: To find the slope of the equation, first we have to find the derivative:

    $$y^2 = 4x^3$$
    $$2y\,dy = 12x^2\,dx$$
    $$\frac{dy}{dx} = \frac{12x^2}{2y}$$
    $$\frac{dy}{dx} = \frac{6x^2}{y}$$

    Now, we can solve for the slope at the given point, $(4, 16)$:

    $$\frac{dy}{dx} = \frac{6x^2}{y}$$
    $$\frac{dy}{dx} = \frac{6(4)^2}{16}$$
    $$\frac{dy}{dx} = \frac{6 \cdot 16}{16}$$
    $$\frac{dy}{dx} = 6$$

    The slope of the graph at $(4, 16)$ is 6.

14. **Question:** At what x value does the equation $y = 7\sqrt{x}$ have a slope of 10?

    Solution: $x = \frac{49}{400}$

Explanation: To solve this problem, first we have to take the derivative:

$$y = 7\sqrt{x}$$
$$y = 7x^{\frac{1}{2}}$$
$$d(y) = d(7x^{\frac{1}{2}})$$
$$dy = \frac{7}{2}x^{\frac{-1}{2}}\,dx$$
$$\frac{dy}{dx} = \frac{7}{2}x^{\frac{-1}{2}}$$
$$\frac{dy}{dx} = \frac{7}{2\sqrt{x}}$$

To find the solution, we set $\frac{dy}{dx}$ to 10 and solve for x:

$$\frac{dy}{dx} = \frac{7}{2\sqrt{x}}$$
$$10 = \frac{7}{2\sqrt{x}}$$
$$\sqrt{x} = \frac{7}{20}$$
$$x = \left(\frac{7}{20}\right)^2$$
$$x = \frac{49}{400}$$

The slope of 10 occurs at $x = \frac{49}{400}$.

15. **Question:** Find the equation of the line tangent to the graph of $x^2 + y^2 = 8$ at $(2, 2)$.

    Solution: $y = -x + 4$

    Explanation: To find the equation of the tangent line, we need a point and a slope. We are given the point $(2, 2)$, so we just need a slope, which is the derivative:

    $$x^2 + y^2 = 8$$
    $$d(x^2 + y^2) = d(8)$$
    $$d(x^2) + d(y^2) = d(8)$$
    $$2x\,dx + 2y\,dy = 0 \qquad 2y\,dy = -2x\,dx \qquad \frac{dy}{dx} = \frac{-2x}{2y}$$
    $$\frac{dy}{dx} = \frac{-x}{y}$$

So, the slope at $(2, 2)$ is:

$$\frac{dy}{dx} = \frac{-x}{y}$$

$$\frac{dy}{dx} = \frac{-2}{2}$$

$$\frac{dy}{dx} = -1$$

Now, we have all parts of the equation of the line except b, which we can solve for using our point, $(2, 2)$:

$$y = -x + b$$

$$2 = -2 + b$$

$$4 = b$$

Therefore, the equation for the line is $y = -x + 4$.

Chapter 11

Differentials of Composite Functions

Find the differential:

1. **Question:** $y = 3x^2 + 6x + 5$

   Solution: $dy = 6x\,dx + 6\,dx$

   Explanation:

   $$y = 3x^2 + 6x + 5$$
   $$d(y) = d(3x^2 + 6x + 5)$$
   $$dy = d(3x^2) + d(6x) + d(5)$$
   $$dy = 6x\,dx + 6\,dx + 0$$
   $$dy = 6x\,dx + 6\,dx$$

2. **Question:** $y = 2^{3x} - 20x$

   Solution: $dy = 3\ln(2)2^{3x}\,dx - 20\,dx$

   Explanation: Start by taking the differential of the original equation as far as you can go normally:

   $$y = 2^{3x} - 20x$$
   $$d(y) = d(2^{3x} - 20x)$$
   $$dy = d(2^{3x}) - d(20x)$$
   $$dy = d(2^{3x}) - 20\,dx$$

   Now, we are stuck on 2^{3x} because the power is a *function* of x and not x itself. So we will make a substitution ($u = 3x$) to convert this to a simple differential that matches the rule: 2^u:

   $$dy = d(2^{3x}) - 20\,dx$$
   $$u = 3x$$
   $$dy = d(2^u) - 20\,dx$$
   $$dy = \ln(2)2^u\,du - 20\,dx$$

   This is great, but we now need to get this back in terms of x. We can substitute $3x$ back in for u, but how do we find du? We can find it by taking the differential of the equation that establishes u:

   Now we can substitute these back into our differential:

   $$dy = \ln(2)2^u\,du - 20\,dx \qquad u = 3x$$
   $$du = 3\,dx$$
   $$dy = \ln(2)2^{3x}3\,dx - 20\,dx$$
   $$dy = 3\ln(2)2^{3x}\,dx - 20\,dx$$

3. **Question:** $y = \sin(x) + 6x^2$

   Solution: $dy = \cos(x)\,dx + 12x\,dx$

   Explanation:

   $$y = \sin(x) + 6x^2$$
   $$d(y) = d(\sin(x) + 6x^2)$$
   $$dy = d(\sin(x)) + d(6x^2)$$
   $$dy = \cos(x)\,dx + 12x\,dx$$

4. **Question:** $y = \sin(3x^5)$

   Solution: $dy = 15x^4 \cos(3x^5)\, dx$

   Explanation: In this equation, we cannot do anything to the right-hand side until we get a u-substitution. Therefore, we will set $u = 3x^5$. Then we can do the problem:

   $$y = \sin(3x^5)$$
   $$u = 3x^5$$
   $$y = \sin(u)$$
   $$dy = \cos(u)\, du$$

   Before we substitute back in, we have to find du:

   $$u = 3x^5$$
   $$du = 15x^4$$

   Now we can substitute back into our equation:

   $$dy = \cos(u)\, du$$
   $$dy = \cos(3x^5) \cdot 15x^4\, dx$$
   $$dy = 15x^4 \cos(3x^5)\, dx$$

5. **Question:** $\sin(x + y) = x^2$

   Solution: $\cos(x + y)(dx + dy) = 2x\, dx$

   Explanation: This is the same process, but our u will be a combination of both x and y:

   $$\sin(x + y) = x^2$$

$d(\sin(x + y)) = d(x^2)$	differentiate
$d(\sin(x + y)) = 2x\, dx$	power rule
$u = x + y$	create substitution
$d(\sin(u)) = 2x\, dx$	apply substitution
$\cos(u)\, du = 2x\, dx$	sine rule
$u = x + y$	find du
$du = dx + dy$	
$\cos(x + y)(dx + dy) = 2x\, dx$	substitute back in

6. **Question:** $y = \sin(3^{x^5})$

   Solution: $dy = 5\ln(3)\cos(3^{x^5})3^{x^5}x^4\, dx$

   Explanation: This problem requires multiple applications of substitution.

   $$y = \sin(3^{x^5})$$
   $$u = 3^{x^5}$$
   $$y = \sin(u)$$
   $$dy = \cos(u)\, du$$

   $$u = 3^{x^5}$$
   $$v = x^5$$
   $$u = 3^v$$
   $$du = \ln(3)3^v\, dv$$

   $$v = x^5$$
   $$dv = 5x^4\, dx$$

   $$dy = \cos(u)\, du$$
   $$dy = \cos(3^{x^5})\ln(3)3^v\, dv$$
   $$dy = \cos(3^{x^5})\ln(3)3^{x^5}5x^4\, dx$$
   $$dy = 5\ln(3)\cos(3^{x^5})3^{x^5}x^4\, dx$$

7. **Question:** $\sin(x - \cos(x^2)) = e^{x-y}$

   Solution: $\cos(x - \cos(x^2))(dx + 2x\sin(x^2)\, dx) = e^{x-y}(dx - dy)$

Explanation:

$$\sin(x - \cos(x^2)) = 2^{x-y}$$

$$d(\sin(x - \cos(x^2))) = d(e^{x-y})$$

$$u = x - \cos(x^2)$$

$$v = x - y$$

$$d(\sin(u)) = d(e^v)$$

$$\cos(u)\, du = e^v\, dv$$

$$u = x - \cos(x^2)$$

$$w = x^2$$

$$u = x - \cos(w)$$

$$du = dx - -\sin(w)\, dw$$

$$du = dx + \sin(w)\, dw$$

$$v = x - y$$

$$dv = dx - dy$$

$$w = x^2$$

$$dw = 2x\, dx$$

$$\cos(u)\, du = e^v\, dv$$

$$\cos(x - \cos(x^2))(dx + \sin(w)\, dw) = e^{x-y}(dx - dy)$$

$$\cos(x - \cos(x^2))(dx + \sin(x^2)2x\, dx) = e^{x-y}(dx - dy)$$

$$\cos(x - \cos(x^2))(dx + 2x\sin(x^2)\, dx) = e^{x-y}(dx - dy)$$

Find the derivative:

8. **Question:** $y = x^3 + 2x^2 + 5x + 10$

   Solution: $\frac{dy}{dx} = 3x^2 + 4x + 5$

   Explanation:

   $$y = x^3 + 2x^2 + 5x + 10$$

   $$dy = 3x^2\, dx + 4x\, dx + 5\, dx$$

   $$\frac{dy}{dx} = 3x^2 + 4x + 5$$

9. **Question:** $y^3 - x^2 = 3x$

   Solution: $\frac{dy}{dx} = \frac{3+2x}{3y^2}$

   Explanation:

   $$y^3 - x^2 = 3x$$

   $$3y^2\, dy - 2x\, dx = 3\, dx$$

   $$3y^2\, dy = 3\, dx + 2x\, dx$$

   $$\frac{dy}{dx} = \frac{3 + 2x}{3y^2}$$

10. **Question:** $\sin(x^2) = y$

    Solution: $2x \cos(x^2) = \frac{dy}{dx}$

    Explanation:

    $$\sin(x^2) = y$$

    $$d(\sin(x^2)) = d(y)$$

    $$u = x^2$$

    $$d(\sin(u)) = dy$$

    $$\cos(u)\, du = dy$$

    $$u = x^2$$

    $$du = 2x\, dx$$

$$\cos(x^2)2x\,dx = dy$$

$$2x\cos(x^2)\,dx = dy$$

$$2x\cos(x^2) = \frac{dy}{dx}$$

$$u = 3^x$$

$$du = \ln(3)3^x\,dx$$

$$-\sin(u)\,du = 10y\,dy$$

$$-\sin(3^x)\ln(3)3^x\,dx = 10y\,dy$$

$$\frac{-\ln(3)\sin(3^x)3^x}{10y} = \frac{dy}{dx}$$

11. **Question:** $\sin(y-2) = \cos(x^2)$

Solution: $\frac{dy}{dx} = \frac{-2x\sin(x^2)}{\cos(y-2)}$

Explanation:

$$\sin(y-2) = \cos(x^2)$$

$$d(\sin(y-2)) = d(\cos(x^2))$$

$$u = y-2$$

$$v = x^2$$

$$d(\sin(u)) = d(\cos(v))$$

$$\cos(u)\,du = -\sin(v)\,dv$$

$$u = y-2$$

$$du = dy$$

$$v = x^2$$

$$dv = 2x\,dx$$

$$\cos(u)\,du = -\sin(v)\,dv$$

$$\cos(y-2)\,dy = -\sin(x^2)2x\,dx$$

$$\cos(y-2)\,dy = -2x\sin(x^2)\,dx$$

$$\frac{dy}{dx} = \frac{-2x\sin(x^2)}{\cos(y-2)}$$

13. **Question:** $y = 3^{x^2+5x}$

Solution: $\frac{dy}{dx} = \ln(3)3^{x^2+5x}(2x+5)$

Explanation:

$$y = 3^{x^2+5x}$$

$$d(y) = d(3^{x^2+5x})$$

$$u = x^2+5x$$

$$d(y) = d(3^u)$$

$$dy = \ln(3)3^u\,du$$

$$u = x^2+5x$$

$$du = 2x\,dx + 5\,dx$$

$$dy = \ln(3)3^{x^2+5x}(2x\,dx + 5\,dx)$$

$$dy = \ln(3)3^{x^2+5x}(2x+5)\,dx$$

$$\frac{dy}{dx} = \ln(3)3^{x^2+5x}(2x+5)$$

12. **Question:** $\cos(3^x) = 5y^2$

Solution: $\frac{dy}{dx} = \frac{-\ln(3)\sin(3^x)3^x}{10y}$

Explanation:

$$\cos(3^x) = 5y^2$$

$$d(\cos(3^x)) = d(5y^2)$$

$$u = 3^x$$

$$d(\cos(u)) = d(5y^2)$$

$$-\sin(u)\,du = 10y\,dy$$

14. **Question:** $\sin(x-3y) = 2^x$

Solution: $\frac{dy}{dx} = \frac{\ln(2)2^x - \cos(x-3y)}{-3\cos(x-3y)}$

Explanation:

$$\sin(x-3y) = 2^x$$

$$d(\sin(x-3y)) = d(2^x)$$

$$u = x-3y$$

$$d(\sin(u)) = d(2^x)$$

$$\cos(u)\,du = \ln(2)2^x\,dx$$

$$u = x - 3y$$

$$du = dx - 3\,dy$$

$$\cos(u)\,du = \ln(2)2^x\,dx$$

$$\cos(x - 3y)(dx - 3\,dy) = \ln(2)2^x\,dx$$

$$\cos(x - 3y)\,dx - 3\cos(x - 3y)\,dy = \ln(2)2^x\,dx$$

$$-3\cos(x - 3y)\,dy = \ln(2)2^x\,dx - \cos(x - 3y)\,dx$$

$$\frac{dy}{dx} = \frac{\ln(2)2^x - \cos(x - 3y)}{-3\cos(x - 3y)}$$

15. **Question:** $\sin(x^3) = x^2 + 5x + \cos(y - 6)$

Solution: $\frac{dy}{dx} = \frac{2x + 5 - 3x^2\,\cos(x^3)}{y - 6}$

Explanation:

$$\sin(x^3) = x^2 + 5x + \cos(y - 6)$$

$$d(\sin(x^3)) = d(x^2 + 5x + \cos(y - 6))$$

$$u = x^3$$

$$v = y - 6$$

$$d(\sin(u)) = d(x^2 + 5x + \cos(v))$$

$$\cos(u)\,du = 2x\,dx + 5\,dx + -\sin(v)\,dv$$

$$du = 3x^2\,dx$$

$$dv = dy$$

$$\cos(x^3)3x^2\,dx = 2x\,dx + 5\,dx + -\sin(y - 6)\,dy$$

$$\sin(y - 6)\,dy = 2x\,dx + 5\,dx - \cos(x^3)3x^2\,dx$$

$$\frac{dy}{dx} = \frac{2x + 5 - 3x^2\,\cos(x^3)}{\sin(y - 6)}$$

Chapter 12

Additional Differential Rules

Find the differential (you don't need to simplify):

1. **Question:** $y = x \cdot 2^x$

   Solution: $dy = \ln(2)\, x\, 2^x\, dx + 2^x\, dx$

   Explanation: This uses the product rule:

   $$y = x \cdot 2^x$$
   $$d(y) = d(x \cdot 2^x)$$
   $$u = x \qquad \text{make substitutions}$$
   $$du = dx$$
   $$v = 2^x$$
   $$dv = \ln(2)2^x\, dx$$
   $$d(y) = d(uv)$$
   $$dy = u\, dv + v\, du \qquad \text{product rule}$$
   $$dy = (x)(\ln(2)2^x\, dx) + (2^x)(dx)$$
   $$dy = \ln(2)\, x\, 2^x\, dx + 2^x\, dx$$
   $$\frac{dy}{dx} = \ln(2)\, x\, 2^x + 2^x$$

2. **Question:** $y^2 = x^3\, 4^x$

   Solution: $2y\, dy = \ln(4)x^3\, 4^x\, dx + 3x^2\, 4^x\, dx$

Explanation:

$$y^2 = x^3\, 4^x$$
$$d(y^2) = d(x^3\, 4^x)$$
$$u = x^3$$
$$du = 3x^2\, dx$$
$$v = 4^x$$
$$dv = \ln(4)4^x\, dx$$
$$d(y^2) = d(uv)$$
$$2y\, dy = u\, dv + v\, du$$
$$2y\, dy = (x^3)(\ln(4)4^x\, dx) + (4^x)(3x^2\, dx)$$
$$2y\, dy = \ln(4)x^3\, 4^x\, dx + 3x^2\, 4^x\, dx$$

3. **Question:** $y^x = 3x^2$

   Solution: $\quad \ln(y)y^x\, dx \ + \ xy^{x-1}\, dy \ = \ 6x\, dx$

   Explanation: This equation uses the

generalized power rule to solve:

$$y^x = 3x^2$$

$$d(y^x) = d(3x^2)$$

$$u = y$$

$$du = dy$$

$$v = x$$

$$dv = dx$$

$$d(u^v) = d(3x^2)$$

$$\ln(u)u^v\,dv + vu^{v-1}\,du = 6x\,dx$$

$$\ln(y)y^x\,dx + xy^{x-1}\,dy = 6x\,dx$$

4. Question: $y = \frac{x^2-3}{x^4+5}$

Solution: $dy = \frac{(x^4+5)(2x\,dx)-(x^2-3)(4x^3\,dx)}{(x^4+5)^2}$

Explanation: This uses the quotient rule:

$$y = \frac{x^2-3}{x^4+5}$$

$$d(y) = d(\frac{x^2-3}{x^4+5})$$

$$u = x^2 - 3$$

$$du = 2x\,dx$$

$$v = x^4 + 5$$

$$dv = 4x^3\,dx$$

$$d(y) = d(\frac{u}{v})$$

$$dy = \frac{v\,du - u\,dv}{v^2}$$

$$dy = \frac{(x^4+5)(2x\,dx) - (x^2-3)(4x^3\,dx)}{(x^4+5)^2}$$

5. Question: $\frac{y}{x} = x^2$

Solution: $\frac{x\,dy - y\,dx}{x^2} = 2x\,dx$

Explanation:

$$\frac{y}{x} = x^2$$

$$d\left(\frac{y}{x}\right) = d(x^2)$$

$$\frac{x\,dy - y\,dx}{x^2} = 2x\,dx$$

6. Question: $y^2 x^2 = e^x$

Solution: $2xy^2\,dx + 2x^2 y\,dy = e^x\,dx$

Explanation:

$$y^2 x^2 = e^x$$

$$d(y^2 x^2) = d(e^x)$$

$$u = y^2$$

$$du = 2y\,dy$$

$$v = x^2$$

$$dv = 2x\,dx$$

$$d(uv) = d(e^x)$$

$$u\,dv + v\,du = e^x\,dx$$

$$(y^2)(2x\,dx) + (x^2)(2y\,dy) = e^x\,dx$$

$$2xy^2\,dx + 2x^2 y\,dy = e^x\,dx$$

7. Question: $x^{y-5x} = 2xy$

Solution: $\ln(x)x^{y-5x}(dy - 5\,dx) + (y - 5x)x^{y-5x-1}\,dx = 2(x\,dy + y\,dx)$

Explanation:

$$x^{y-5x} = 2xy$$

$$d(x^{y-5x}) = d(2xy)$$

$$u = y - 5x$$

$$du = dy - 5\,dx$$

$$d(x^u) = 2d(xy)$$

$$\ln(x)x^u\,du + ux^{u-1}\,dx = 2(x\,dy + y\,dx)$$

$$\ln(x)x^{y-5x}(dy - 5\,dx) + (y - 5x)x^{y-5x-1}\,dx = 2(x\,dy + y\,dx)$$

the right.

$$x^{x^2} = xy$$

$$d(x^{x^2}) = d(xy)$$

$$u = x^2$$

$$du = 2x \, dx$$

$$d(x^u) = d(xy)$$

$$\ln(x)x^u \, du + ux^{u-1} \, dx = x \, dy + y \, dx$$

$$(\ln(x)x^{x^2})(2x \, dx) + (x^2)(x^{x^2-1}) \, dx = x \, dy + y \, dx$$

$$(\ln(x)x^{x^2})(2x \, dx) + (x^2)(x^{x^2-1}) \, dx - y \, dx = x \, dy$$

$$\frac{(\ln(x)x^{x^2})(2x) + (x^2)(x^{x^2-1}) - y}{x} = \frac{dy}{dx}$$

$$\frac{2\ln(x)x^{x^2+1} + x^{x^2+1} - y}{x} = \frac{dy}{dx}$$

Note that the simplification in the last step is based on exponent rules. For instance, $x \cdot x^{x^2} = x^{x^2+1}$. Likewise, $x^2 \cdot x^{x^2-1} = x^{x^2+1}$.

You can actually simplify it further by separating out the pieces of the fraction:

$$\frac{dy}{dx} = \frac{2\ln(x)x^{x^2+1} + x^{x^2+1} - y}{x}$$

$$\frac{dy}{dx} = \frac{2\ln(x)x^{x^2+1}}{x} + \frac{x^{x^2+1}}{x} - \frac{y}{x}$$

$$\frac{dy}{dx} = 2\ln(x)x^{x^2} + x^{x^2} - \frac{y}{x}$$

$$\frac{dy}{dx} = (2\ln(x) + 1)x^{x^2} - \frac{y}{x}$$

8. **Question:** $3xy^2 = 5$

   Solution: $6xy \, dy + 3y^2 \, dx = 0$

   Explanation: In this problem we will use the constant multiplier rule to make the use of the product rule simpler:

   $$3xy^2 = 5$$

   $$d(3xy^2) = d(5)$$

   $$3 \, d(xy^2) = 0$$

   $$u = y^2$$

   $$du = 2y \, dy$$

   $$3 \, d(xu) = 0$$

   $$3(x \, du + u \, dx) = 0$$

   $$3(x \, 2y \, dy + y^2 \, dx) = 0$$

   $$6xy \, dy + 3y^2 \, dx = 0$$

Find the derivative:

9. **Question:** $x^{x^2} = xy$

   Solution: $\frac{dy}{dx} = (2\ln(x) + 1)x^{x^2} - \frac{y}{x}$

   Explanation: Here we will use the generalized power rule on the left, and the product rule on

10. **Question:** $y^x = 3x^2 + 6x$

    Solution: $\frac{dy}{dx} = \frac{6x \, dx + 6 - \ln(y)y^x}{xy^{x-1}}$

    Explanation:

    $$y^x = 3x^2 + 6x$$

    $$\ln(y)y^x \, dx + xy^{x-1} \, dy = 6x \, dx + 6 \, dx$$

    $$xy^{x-1} \, dy = 6x \, dx + 6 \, dx - \ln(y)y^x \, dx$$

    $$\frac{dy}{dx} = \frac{6x \, dx + 6 - \ln(y)y^x}{xy^{x-1}}$$

Use the derivative tables in Appendix H.6 to perform the following derivatives:

11. **Question:** $y = \sec(x^2)$

    Solution: $\frac{dy}{dx} = 2x\,\sec(x^2)\tan(x^2)$

    Explanation: To find the differential, look up the rule for $d(\sec(u))$ in the appendix. You will find that $d(\sec(u)) = \sec(u)\tan(u)\,du$.

    $$y = \sec(x^2)$$
    $$d(y) = d(\sec(x^2))$$
    $$u = x^2$$
    $$du = 2x\,dx$$
    $$d(y) = d(\sec(u))$$
    $$dy = \sec(u)\tan(u)\,du$$
    $$dy = \sec(x^2)\tan(x^2)(2x\,dx)$$
    $$dy = 2x\,\sec(x^2)\tan(x^2)\,dx$$
    $$\frac{dy}{dx} = 2x\,\sec(x^2)\tan(x^2)$$

12. **Question:** $\ln(y) = \cos(x)$

    Solution: $\frac{dy}{dx} = -y\,\sin(x)$

    Explanation:

    $$\ln(y) = \cos(x)$$
    $$d(\ln(y)) = d(\cos(x))$$
    $$\frac{dy}{y} = -\sin(x)\,dx \qquad \frac{dy}{dx} = -y\,\sin(x)$$

13. **Question:** $y = \cot(\ln(x))$

    Solution: $\frac{dy}{dx} = \frac{-\csc^2(\ln(x))}{x}$

Explanation:

$$y = \cot(\ln(x))$$
$$d(y) = d(\cot(\ln(x)))$$
$$u = \ln(x)$$
$$du = \frac{dx}{x}$$
$$d(y) = d(\cot(u))$$
$$dy = -\csc^2(u)\,du$$
$$dy = -\csc^2(\ln(x))\left(\frac{dx}{x}\right)$$
$$dy = \frac{-\csc^2(\ln(x))}{x}\,dx$$
$$\frac{dy}{dx} = \frac{-\csc^2(\ln(x))}{x}$$

14. **Question:** $\sin(y) = \log_4(x^2)$

    Solution: $\frac{dy}{dx} = \frac{2}{\ln(4)x\,\cos(y)}$

    Explanation: This problem uses not just the log rule, but the log rule for a *specific* base. Note that there is a log rule for base n. Use that log rule to solve.

    $$\sin(y) = \log_4(x^2)$$
    $$d(\sin(y)) = d(\log_4(x^2))$$
    $$u = x^2$$
    $$du = 2x\,dx$$
    $$d(\sin(y)) = d(\log_4(u))$$
    $$\cos(y)\,dy = \frac{du}{\ln(4)u}$$
    $$\cos(y)\,dy = \frac{2x\,dx}{\ln(4)x^2}$$
    $$\cos(y)\,dy = \frac{2\,dx}{\ln(4)x}$$
    $$\frac{dy}{dx} = \frac{2}{\ln(4)x\,\cos(y)}$$

15. **Question:** $y^2 = \sec(x^2 + \ln(x))$

    Solution: $\frac{dy}{dx} = \frac{\sec(x^2+\ln(x))\tan(x^2+\ln(x))\left(2x+\frac{1}{x}\right)}{2y}$

Explanation:

$$y^2 = \sec(x^2 + \ln(x))$$

$$d(y^2) = d(\sec(x^2 + \ln(x)))$$

$$u = x^2 + \ln(x)$$

$$du = 2x\,dx + \frac{dx}{x}$$

$$d(y^2) = d(\sec(u))$$

$$2y\,dy = \sec(u)\tan(u)\,du$$

$$2y\,dy = \sec(x^2 + \ln(x))\tan(x^2 + \ln(x))\left(2x\,dx + \frac{dx}{x}\right)$$

$$\frac{dy}{dx} = \frac{\sec(x^2 + \ln(x))\tan(x^2 + \ln(x))\left(2x + \frac{1}{x}\right)}{2y}$$

16. **Question:** $xy = \arccos(x^2)$

Solution: $\frac{dy}{dx} = \frac{\frac{-2x}{\sqrt{1-x^4}} - y}{x}$

Explanation: In this case, we are using the arccos() function (other books may call this $\cos^{-1}()$, but we use *arccos* to prevent confusing inverse functions and negative exponents). The differential rule for this makes use of a constant divisor n. Since we don't have a constant divisor, then we will just use 1 for n.

$$xy = \arccos(x^2)$$

$$d(xy) = d(\arccos(x^2))$$

$$u = x^2$$

$$du = 2x\,dx$$

$$d(xy) = d(\arccos(u))$$

$$x\,dy + y\,dx = \frac{-du}{\sqrt{1-u^2}}$$

$$x\,dy + y\,dx = \frac{-2x\,dx}{\sqrt{1-(x^2)^2}}$$

$$x\,dy + y\,dx = \frac{-2x\,dx}{\sqrt{1-x^4}}$$

$$x\,dy = \frac{-2x\,dx}{\sqrt{1-x^4}} - y\,dx$$

$$\frac{dy}{dx} = \frac{\frac{-2x}{\sqrt{1-x^4}} - y}{x}$$

17. **Question:** $\sin(x) = \log_y(x^2)$

Solution: $\frac{dy}{dx} = \frac{(\ln(y)\,x\,\cos(x) - 2)(y\,\ln(y))}{x\,\ln(x^2)}$

Explanation: Note that this rule uses the logarithm rule that has a *variable* base, and not a constant one. In the table of differentials, constants are listed as n or C. Therefore, look for the one listed as $d(\log_u(v))$, not the one listed as $d(\log_n(v))$. Note that there are two listed—you can choose either one as they are equivalent.

$$\sin(x) = \log_y(x^2)$$

$$d(\sin(x)) = d(\log_y(x^2))$$

$$u = x^2$$

$$du = 2x\,dx$$

$$d(\sin(x)) = d(\log_y(u))$$

$$\cos(x)\,dx = \frac{du}{\ln(y)u} - \frac{\ln(u)dy}{y\,(\ln(y))^2}$$

$$\cos(x)\,dx = \frac{2x\,dx}{\ln(y)x^2} - \frac{\ln(x^2)dy}{y\,(\ln(y))^2}$$

$$\cos(x)\,dx - \frac{2x\,dx}{\ln(y)x^2} = -\frac{\ln(x^2)dy}{y\,(\ln(y))^2}$$

$$\frac{\ln(y)\,x^2\cos(x) - 2x}{\ln(y)x^2}\,dx = -\frac{\ln(x^2)}{y\,(\ln(y))^2}\,dy$$

$$\frac{(\ln(y)\,x^2\cos(x) - 2x)(y\,(\ln(y))^2)}{(\ln(y)x^2)(\ln(x^2))} = \frac{dy}{dx}$$

$$\frac{(\ln(y)\,x\,\cos(x) - 2)(y\,\ln(y))}{x\,\ln(x^2)} = \frac{dy}{dx}$$

18. **Question:** $\frac{\tan(y^2)}{x^3} = 5$

Solution: $\frac{dy}{dx} = \frac{3x^2\tan(y^2)}{2x^3\,y\,\sec^2(y^2)}$ or $\frac{dy}{dx} = \frac{15x^2}{2y\,\sec^2(y^2)}$

Explanation:

$$\frac{\tan(y^2)}{x^3} = 5$$

$$d\left(\frac{\tan(y^2)}{x^3}\right) = d(5)$$

$$u = \tan(y^2)$$

$$du = d(\tan(y^2))$$

$$w = y^2$$

$$dw = 2y\,dy$$

$$du = d(\tan(w))$$

$$du = \sec^2(w)\,dw$$

$$du = 2y\,\sec^2(y^2)\,dy$$

$$v = x^3$$

$$dv = 3x^2\,dx$$

$$d\left(\frac{u}{v}\right) = d(5)$$

$$\frac{v\,du - u\,dv}{v^2} = 0$$

Since this is equal to zero, we can multiply both sides by v^2 to simplify things a bit:

$$v\,du - u\,dv = 0$$

Now we can substitute back in for u and v:

$$(x^3)(2y\,\sec^2(y^2)\,dy) - \tan(y^2)(3x^2\,dx) = 0$$

Now we just need to separate out our dy and dx terms and solve for $\frac{dy}{dx}$:

$$(x^3)(2y\,\sec^2(y^2)\,dy) = \tan(y^2)(3x^2\,dx)$$

$$\frac{dy}{dx} = \frac{\tan(y^2)(3x^2)}{2x^3\,y\,\sec^2(y^2)}$$

$$\frac{dy}{dx} = \frac{3x^2\,\tan(y^2)}{2x^3\,y\,\sec^2(y^2)}$$

Also note that if you wanted to simplify it first, you could have multiplied both sides by x^3 to remove the quotient. This would have led to a slightly different result, but the result would be equivalent over the domain of the original

function:

$$\frac{\tan(y^2)}{x^3} = 5$$

$$\tan(y^2) = 5x^3$$

$$d(\tan(y^2)) = d(5x^3)$$

$$2y\,\sec^2(y^2)\,dy = 15x^2\,dx$$

$$\frac{dy}{dx} = \frac{15x^2}{2y\,\sec^2(y^2)}$$

You can see that this is equivalent to the previous answer if you realize that, in the original equation, $x^3 = \frac{\tan(y^2)}{5}$. If you perform this substitution in the previous answer, it will result in this answer.

This sometimes happens on differentials where the variables intermix on both sides. Depending on your process you might produce different results, but the results are equivalent within the domain of the original function.

19. **Question:** Find the minimum value (both x and y) of the equation $y = 3^{x^2+x}$. You can assume that the minima occurs at the only local extremum.

Solution: The minimum value occurs at $(0.5, 0.7598)$.

Explanation: To find the minima for the function, we need to take the derivative and set the derivative to zero:

$$y = 3^{x^2+x}$$

$$d(y) = d(3^{x^2+x})$$

$$u = x^2 + x$$

$$du = 2x\,dx + dx$$

$$d(y) = d(3^u)$$

$$dy = \ln(3)3^u\,du$$

$$dy = \ln(3)3^{x^2+x}(2x\,dx + dx)$$

$$dy = \ln(3)3^{x^2+x}(2x + 1)\,dx$$

$$\frac{dy}{dx} = \ln(3)3^{x^2+x}(2x + 1)$$

Now that we have the derivative, we solve for where the derivative is 0. Remember that dividing zero by anything leaves us with zero.

$$\frac{dy}{dx} = \ln(3)3^{x^2+x}(2x+1) \quad \text{the original derivative}$$

$$0 = \ln(3)3^{x^2+x}(2x+1)$$

$$0 = 2x + 1 \qquad \text{divide both sides by the scary terms}$$

$$2x = -1 \qquad \text{solve for } x$$

$$x = \frac{-1}{2}$$

The extrema occurs at $\frac{-1}{2}$. The problem tells you that this will be the global minimum. The y value can be solved by substituting it back in to the original equation:

$$y = 3^{(\frac{-1}{2})^2 + \frac{-1}{2}}$$

$$y = 3^{\frac{1}{4} + \frac{-2}{4}}$$

$$y = 3^{\frac{-1}{4}}$$

$$y = \frac{1}{\sqrt[4]{3}}$$

$$\approx 0.7598$$

Therefore, the minimum value occurs at $(0.5, 0.7598)$.

Chapter 13

Multivariable Differentials

Find the differential:

1. **Question:** $z = x + y$

   Solution: $dz = dx + dy$

   Explanation:

   $$z = x + y$$
   $$d(z) = d(x + y)$$
   $$d(z) = d(x) + d(y)$$
   $$dz = dx + dy$$

2. **Question:** $m^2 = 4q + \sin(b)$

   Solution: $2m\,dm = 4\,dq + \cos(b)\,db$

   Explanation:

   $$m^2 = 4q + \sin(b)$$
   $$d(m^2) = d(4q + \sin(b))$$
   $$d(m^2) = d(4q) + d(\sin(b))$$
   $$2m\,dm = 4\,dq + \cos(b)\,db$$

3. **Question:** $\sin(xy) = z^2$

   Solution: $\cos(xy)(x\,dy + y\,dx) = 2z\,dz$

Explanation:

$$\sin(xy) = z^2$$
$$d(\sin(xy)) = d(z^2)$$
$$u = xy$$
$$du = x\,dy + y\,dx$$
$$d(\sin(u)) = d(z^2)$$
$$\cos(u)\,du = 2z\,dz$$
$$\cos(xy)(x\,dy + y\,dx) = 2z\,dz$$

4. **Question:** $jkb = r^3$

   Solution: $jk\,db + jb\,dk + kb\,dj = 3r^2\,dr$

   Explanation: Here, we will have to break about jkb into separate multiplications in order to use th product rule:

   $$jkb = r^3$$
   $$d(jkb) = d(r^3)$$
   $$d(j(kb)) = d(r^3)$$
   $$u = kb$$
   $$du = k\,db + b\,dk$$
   $$d(ju) = d(r^3)$$
   $$j\,du + u\,dj = 3r^2\,dr$$
   $$j(k\,db + b\,dk) + kb\,dj = 3r^2\,dr$$
   $$jk\,db + jb\,dk + kb\,dj = 3r^2\,dr$$

Find the derivative:

5. **Question:** Find $\frac{dy}{dx}$ for $\sin(y) = x^2$.

   Solution: $\frac{dy}{dx} = \frac{2x}{\cos(y)}$

   Explanation:

   $$\sin(y) = x^2$$
   $$d(\sin(y)) = d(x^2)$$
   $$\cos(y)\,dy = 2x\,dy$$
   $$\frac{dy}{dx} = \frac{2x}{\cos(y)}$$

6. **Question:** Find the derivative of w with respect to y for $\cos(w) = x^2 + y^2$.

   Solution: $\frac{dw}{dy} = -\frac{2x+2y}{\sin(w)}$

   Explanation:

   $$\cos(w) = x^2 + y^2$$
   $$d(\cos(w)) = d(x^2 + y^2)$$
   $$d(\cos(w)) = d(x^2) + d(y^2)$$
   $$-\sin(w)\,dw = 2x\,dx + 2y\,dy$$
   $$\frac{dw}{dy} = -\frac{2x + 2y}{\sin(w)}$$

7. **Question:** Find the derivative of q with respect to m for $b^2 - c^2 = mq$.

   Solution: $\frac{dq}{dm} = 2b\frac{db}{dm} - 2c\frac{dc}{dm} - q$

Explanation:

$$b^2 - c^2 = mq$$
$$d(b^2 - c^2) = d(mq)$$
$$d(b^2) - d(c^2) = d(mq)$$
$$2b\,db - 2c\,dc = m\,dq + q\,dm$$
$$m\,dq = 2b\,db - 2c\,dc - q\,dm$$
$$\frac{dq}{dm} = \frac{2b\,db - 2c\,dc - q\,dm}{dm}$$
$$\frac{dq}{dm} = \frac{2b\,db}{dm} - \frac{2c\,dc}{dm} - \frac{q\,dm}{dm}$$
$$\frac{dq}{dm} = 2b\frac{db}{dm} - 2c\frac{dc}{dm} - q$$

8. **Question:** Find the derivative $\frac{dh}{dk}$ for $\frac{k}{h} = \sin(hq)$.

   Solution: $\frac{dh}{dk} = \frac{h}{k+h^2q\cos(hq)} - \frac{h^3\cos(hq)}{k+h^2q\cos(hq)}\frac{dq}{dk}$

   Explanation:

   $$\frac{k}{h} = \sin(hq)$$
   $$d\left(\frac{k}{h}\right) = d(\sin(hq))$$
   $$\frac{h\,dk - k\,dh}{h^2} = \cos(hq)(h\,dq + q\,dh)$$
   $$h\,dk - k\,dh = h^2\cos(hq)(h\,dq + q\,dh)$$
   $$h\,dk - k\,dh = h^3\cos(hq)\,dq + h^2q\cos(hq)\,dh$$
   $$-k\,dh - h^2q\cos(hq)\,dh = h^3\cos(hq)\,dq - h\,dk$$
   $$k\,dh + h^2q\cos(hq)\,dh = h\,dk - h^3\cos(hq)\,dq$$

   $$\frac{dh}{dk} = \frac{h}{k + h^2q\cos(hq)} - \frac{h^3\cos(hq)}{k + h^2q\cos(hq)}\frac{dq}{dk}$$

9. **Question:** Find the derivative of r with respect to v for $3r^5 = 5x^2$.

   Solution: $\frac{dr}{dv} = \frac{10x}{15r^4}\frac{dx}{dv}$

   Explanation: This is a problem where we are taking the derivative with respect to a variable that doesn't exist in the equation (v). However, the process is just the same:

$$3r^5 = 5x^2$$

$$d(3r^5) = d(5x^2)$$

$$15r^4\,dr = 10x\,dx$$

$$dr = \frac{10x}{15r^4}\,dx$$

$$\frac{dr}{dv} = \frac{10x}{15r^4}\frac{dx}{dv}$$

Note that the last step we simply divided by dv, which is fine since we did it to both sides.

10. **Question:** Find the derivative $\frac{dy}{dt}$ for $\sin(xy) = r^3$.

Solution: $\frac{dy}{dt} = \frac{3r^2}{x\cos(xy)}\frac{dr}{dt} - \frac{y}{x}\frac{dx}{dt}$

Explanation:

$$\sin(xy) = r^3$$

$$d(\sin(xy)) = d(r^3)$$

$$u = xy$$

$$du = x\,dy + y\,dx$$

$$d(\sin(u)) = d(r^3)$$

$$\cos(u)\,du = 3r^2\,dr$$

$$\cos(xy)(x\,dy + y\,dx) = 3r^2\,dr$$

$$x\cos(xy)\,dy + y\cos(xy)\,dx = 3r^2\,dr$$

$$x\cos(xy)\,dy = 3r^2\,dr - y\cos(xy)\,dx$$

$$\frac{dy}{dt} = \frac{3r^2}{x\cos(xy)}\frac{dr}{dt} - \frac{y}{x}\frac{dx}{dt}$$

In the last step we just divided by dt even though it didn't exist in the equation. Since we did the same thing to both sides it works out fine.

11. **Question:** Find the second derivative of y with respect to x of $3y = x^3$.

Solution: The second derivative fo y with respect to x is $2x$.

Explanation: To find the second derivative, we start by finding the first derivative:

$$3y = x^3$$

$$d(3y) = d(x^3)$$

$$3\,dy = 3x^2\,dx$$

$$\frac{dy}{dx} = x^2$$

To find the second derivative, create a new variable (we will use q) and set it to be the first derivative:

$$q = \frac{dy}{dx} = x^2$$

Now, find the derivative of q with respect to x:

$$q = x^2$$

$$d(q) = d(x^2)$$

$$dq = 2x\,dx$$

$$\frac{dq}{dx} = 2x$$

The second derivative fo y with respect to x is $2x$.

12. **Question:** Find the second derivative of h with respect to g for $e^g = \sin(h)$.

Solution: The second derivative of h with respect to g is $\frac{e^g}{\cos(h)} + \frac{e^{2g}\sin(h)}{(\cos(h))^3}$

Explanation: The first step to finding the second derivative is finding the first derivative:

$$e^g = \sin(h)$$

$$d(e^g) = d(\sin(h))$$

$$e^g\,dg = \cos(h)\,dh$$

$$\frac{dh}{dg} = \frac{e^g}{\cos(h)}$$

Now, to find the second derivative, we just need to create a new variable (we will use p) and set it to be the first derivative:

$$p = \frac{dh}{dg} = \frac{e^g}{\cos(h)}$$

Now we will solve for the derivative of p with

respect to g:

$$p = \frac{e^g}{\cos(h)}$$

$$d(p) = d\left(\frac{e^g}{\cos(h)}\right)$$

$$dp = \frac{\cos(h)\,d(e^g) - e^g\,d(\cos(h))}{(\cos(h))^2}$$

$$dp, = \frac{\cos(h)\,e^g\,dg - e^g(-\sin(h))\,dh}{(\cos(h))^2}$$

$$dp = \frac{\cos(h)\,e^g}{(\cos(h))^2}\,dg + \frac{e^g\,\sin(h)}{(\cos(h))^2}\,dh$$

$$\frac{dp}{dg} = \frac{\cos(h)\,e^g}{(\cos(h))^2} + \frac{e^g\,\sin(h)}{(\cos(h))^2}\,\frac{dh}{dg}$$

We could end it there, but notice the term $\frac{dh}{dg}$. That's just the first derivative! Therefore, we can replace $\frac{dh}{dg}$ with the first derivative. This yields:

$$\frac{dp}{dg} = \frac{\cos(h)\,e^g}{(\cos(h))^2} + \frac{e^g\,\sin(h)}{(\cos(h))^2}\,\frac{e^g}{\cos(h)}$$

$$\frac{dp}{dg} = \frac{\cos(h)\,e^g}{(\cos(h))^2} + \frac{(e^g)^2\,\sin(h)}{(\cos(h))^3}$$

$$\frac{dp}{dg} = \frac{e^g}{\cos(h)} + \frac{e^{2g}\,\sin(h)}{(\cos(h))^3}$$

This is the second derivative of h with respect to g.

13. **Question:** Find the third derivative of y with respect to x for $y = x\sin(x)$.

Solution: The third derivative of y with respect to x is $-x\cos(x) - 3\sin(x)$.

Explanation: To find the third derivative, first we need to find the first derivative:

$$y = x\sin(x)$$

$$d(y) = d(x\sin(x))$$

$$dy = x\cos(x)\,dx + \sin(x)\,dx$$

$$\frac{dy}{dx} = x\cos(x) + \sin(x)$$

Next we will set $q = \frac{dy}{dx} = x\cos(x) + \sin(x)$. Now

we will find the second derivative, $\frac{dq}{dx}$:

$$q = x\cos(x) + \sin(x)$$

$$d(q) = d(x\cos(x) + \sin(x))$$

$$dq = -x\sin(x)\,dx + \cos(x)\,dx + \cos(x)\,dx$$

$$dq = -x\sin(x)\,dx + 2\cos(x)\,dx$$

$$\frac{dq}{dx} = -x\sin(x) + 2\cos(x)$$

Next we will set $r = \frac{dq}{dx} = -x\sin(x) + 2\cos(x)$. Now we will find the third derivative, $\frac{dr}{dx}$:

$$r = -x\sin(x) + 2\cos(x)$$

$$d(r) = d(-x\sin(x) + 2\cos(x))$$

$$dr = -x\cos(x)\,dx + -\sin(x)\,dx + -2\sin(x)\,dx$$

$$dr = -x\cos(x)\,dx + -3\sin(x)\,dx$$

$$\frac{dr}{dx} = -x\cos(x) - 3\sin(x)$$

And that is the third derivative.

Chapter 14

Relating Rates Using Differentials

Solve the following problems:

1. **Question:** Imagine you have a trough shaped like an extruded isosceles triangle (with the angle at the bottom) that you are filling with a liquid. The trough has a 90° angle at the vertex, and is 20 centimeters long. Find an equation that relates the height of the liquid in the trough with the volume of the liquid in the trough. Use the variable h to represent the height of the liquid, and the variable v to represent the volume.

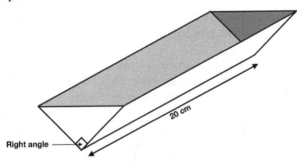

Solution: $V = 20\text{cm} \cdot H^2$

Explanation: The volume of any extruded shape is merely the area of the base mulitiplied by the length of the extrusion (usually, we think about an extruded height, but in this case it is the length). The base is simply a triangle, so we will need to know the triangle's area. Now, the actual dimensions of the triangle will change as the water level rises. However, they will all be similar triangles, because the angles will not change.

The area of a triangle is $\frac{1}{2}$base · height. We will refer to the base as B and the height as H. The height will be a variable (since the

problem wants the result to relate the height to the volume). Therefore, we need to find an equation that relates the length of the base to the height.

If we draw a straight line down the triangle, we get two right triangles, with an angle of $\frac{\pi}{4}$ radians (i.e, 45°).

The tangent of that angle will give us the ratio of half the base to the height. In other words:

$$\frac{1}{2}B = \tan(\frac{\pi}{4})H$$

$$\frac{1}{2}B = 1 \cdot H$$

$$B = 2H$$

Therefore, the area of the triangle will be:

$$\frac{1}{2}B \cdot H = \frac{1}{2}(2H)H = H^2$$

Since the trough is 20cm long, then the volume of the trough is:

$$V = 20\text{cm} \cdot H^2$$

2. **Question:** Find the differential of the previous equation.

Solution: $dV = 40\text{cm}\, H\, dH$

3. **Question:** Divide the entirety of the previous equation by dt (the differential of time).

Solution: $\frac{dV}{dt} = 40\text{cm}\, H\, \frac{dH}{dt}$

4. **Question:** Solve the previous equation for the change in height with respect to time.

Solution: $\frac{dH}{dt} = \frac{1}{40\text{cm } H} \frac{dV}{dt}$

Explanation:

$$\frac{dV}{dt} = 40\text{cm } H \frac{dH}{dt}$$

$$\frac{dH}{dt} = \frac{1}{40\text{cm } H} \frac{dV}{dt}$$

5. **Question:** If you are filling the trough from empty at a rate of 5 milliliters per minute (1 milliliter equals 1 cubic centimeter), after 16 minutes, what is the rate of change of the height of the liquid?

Solution: $0.0625 \frac{\text{cm}}{\text{min}}$

Explanation: To solve this, we should recognize the we are getting $5 \frac{\text{cm}^3}{\text{min}}$ of liquid. This is $\frac{dV}{dt}$. The other value we need is the current height. We can solve from that from the *total* volume. The total volume will simply be the rate of volume increase multiplied by the amount of time, or $5 \frac{\text{cm}^3}{\text{min}} \cdot 16 \text{ min} = 80 \text{ cm}^3$. Therefore, we can use the equation relating the height and the volume from an earlier question to solve for the height at this point:

$$V = 20\text{cm } H^2$$

$$80 \text{ cm}^3 = 20 \text{ cm } H^2$$

$$\frac{80 \text{ cm}^3}{20 \text{ cm}} = H^2$$

$$4 \text{ cm}^2 = H^2$$

$$2 \text{ cm} = H$$

Therefore, the height is currently 2 cm. We can plug this back into our equation for $\frac{dH}{dt}$ and find

the current rate of change for this:

$$\frac{dH}{dt} = \frac{1}{40\text{cm } H} \frac{dV}{dt}$$

$$\frac{dH}{dt} = \frac{1}{40\text{cm } 2\text{ cm}} 5 \frac{\text{cm}^3}{\text{min}}$$

$$\frac{dH}{dt} = \frac{1}{80\text{cm}^2} 5 \frac{\text{cm}^3}{\text{min}}$$

$$\frac{dH}{dt} = \frac{1}{16} \frac{\text{cm}}{\text{min}}$$

$$\frac{dH}{dt} = \frac{1}{16} \frac{\text{cm}}{\text{min}}$$

$$\frac{dH}{dt} = 0.0625 \frac{\text{cm}}{\text{min}}$$

6. **Question:** Two cars start at the same point. One travels South at 60 miles per hour and the other travels East at 15 miles per hour. After two hours, what rate is the distance between the cars increasing?

Solution: The distance between the cars is increasing at a rate of $61.8466 \frac{\text{miles}}{\text{hr}}$

Explanation: If you imagine a car traveling South and another East, what shape are the cars forming? They are formin a right triangle. The distance between them is the hypotenuse of that triangle. Therefore, we are going to use the Pythagorean theorem to talk about the distances. D will be the distance between the cars, S will be the distance travelled South, and E will be the distance travelled East. We will also want to convert this to a related rate equation.

$$D^2 = S^2 + E^2$$

$$2D \, dD = 2S \, dS + 2E \, dE$$

$$D \, dD = S \, dS + E \, dE$$

$$D \frac{dD}{dt} = S \frac{dS}{dt} + E \frac{dE}{dt}$$

So, $\frac{dS}{dt}$ is the speed that the car traveling South is traveling, which is $60 \frac{\text{miles}}{\text{hr}}$. Likewise, $\frac{dE}{dt}$ will be $15 \frac{\text{miles}}{\text{hr}}$. Now we just need the actual values of S, D, and E.

This is easy enough because we have both the rate and the time, so we can just multiply them together.

$$S = 60 \, \frac{\text{miles}}{\text{hr}} \cdot 2 \, \text{hr} = 120 \, \text{miles}$$

$$E = 15 \, \frac{\text{miles}}{\text{hr}} \cdot 2 \, \text{hr} = 30 \, \text{miles}$$

$$D^2 = S^2 + E^2 = (120 \, \text{miles})^2 + (30 \, \text{miles})^2$$

$$D^2 = 15300 \, \text{miles}^2$$

$$D = \sqrt{15300 \, \text{miles}^2} \approx 123.6932 \, \text{miles}$$

Now we can throw all of these into our related rate equation, and solve for $\frac{dD}{dt}$:

$$D \frac{dD}{dt} = S \frac{dS}{dt} + E \frac{dE}{dt}$$

$$\frac{dD}{dt} = \frac{S \frac{dS}{dt} + E \frac{dE}{dt}}{D}$$

$$\frac{dD}{dt} = \frac{120 \, \text{miles} \cdot 60 \, \frac{\text{miles}}{\text{hr}} + 30 \, \text{miles} \cdot 15 \, \frac{\text{miles}}{\text{hr}}}{123.6932 \, \text{miles}}$$

$$\frac{dD}{dt} = \frac{7200 \, \frac{\text{miles}^2}{\text{hr}} + 450 \, \frac{\text{miles}^2}{\text{hr}}}{123.6932 \, \text{miles}}$$

$$\frac{dD}{dt} = \frac{7650 \, \frac{\text{miles}^2}{\text{hr}}}{123.6932 \, \text{miles}}$$

$$\frac{dD}{dt} = 61.8466 \, \frac{\text{miles}}{\text{hr}}$$

7. **Question:** Look at the example in Section 14.3. Calculate the rate of change of the *angle* between the ladder and the ground after a half a second of pushing. Remember (from Section 11.1) to use radians for angle measures, but also remember that radians are an implied unit, and do not normally have a unit name attached (i.e., instead of saying $1 \, \frac{\text{radian}}{\text{sec}}$ you would normally just say $\frac{1}{\text{sec}}$ for one radian-per-second). Therefore, you don't need to keep track of the radian unit during your calculation (and, in fact, it might lead to confusing results if you tried).

Solution: The angle between the ladder and the ground is changing at a rate of $\frac{-0.1952}{\text{sec}}$.

Explanation: In the example, B was for the base of the triangle (i.e., the ground between the ladder and the house), H was for the height from the bottom of the house to wherever the ladder touched the house, and L is for the length of the ladder. The angle we are interested in is the angle between the ground and the ladder, which we will call θ. So, first, what we need is a static definition for the angle.

Remember that in a right triangle the cosine of an angle is the adjacent divided by the hypotenuse. Therefore, we can say that:

$$\cos(\theta) = \frac{B}{L}$$

This is our static relationship. Now we need to convert this to a dynamic relationship using the differential, and then to a related rate equation. Remember that we can treat L as a constant, since the length of the ladder does not change:

$$\cos(\theta) = \frac{B}{L}$$

$$-\sin(\theta) \, d\theta = \frac{dB}{L}$$

$$-\sin(\theta) \frac{d\theta}{dt} = \frac{1}{L} \frac{dB}{dt}$$

Now we need to solve for $\frac{d\theta}{dt}$:

$$-\sin(\theta) \frac{d\theta}{dt} = \frac{1}{L} \frac{dB}{dt}$$

$$\frac{d\theta}{dt} = \frac{-1}{L \sin(\theta)} \frac{dB}{dt}$$

We know L (the length of the ladder—this is $11 \, \text{ft}$), $\frac{dB}{dt}$ (from the problem—this is $2 \, \frac{\text{ft}}{\text{sec}}$), so all we need to solve this is θ. In the original example, we calculated that, after a half a second, the base would extend out to $4 \, \text{ft}$. Therefore, we can solve for θ as follows:

$$\cos(\theta) = \frac{B}{L}$$

$$\cos(\theta) = \frac{4 \, \text{ft}}{11 \, \text{ft}}$$

$$\cos(\theta) = \frac{4}{11}$$

$$\theta = \arccos\left(\frac{4}{11}\right)$$

$$\theta \approx 1.1986$$

Therefore, θ is 1.1986 radians at this point (which is about 69° if you would like to visualize it better). Now that we have θ we have everything we need to solve for $\frac{d\theta}{dt}$:

$$\frac{d\theta}{dt} = \frac{-1}{L \, \sin(\theta)} \frac{dB}{dt}$$

$$\frac{d\theta}{dt} = \frac{-1}{11 \, \text{ft} \, \sin(1.1986)} \cdot 2 \, \frac{\text{ft}}{\text{sec}}$$

$$\frac{d\theta}{dt} = \frac{-2}{11 \, \sin(1.1986) \, \text{sec}}$$

$$\frac{d\theta}{dt} = \frac{-2}{10.2465 \, \text{sec}}$$

$$\frac{d\theta}{dt} = \frac{-0.1952}{\text{sec}}$$

Note that there was no resulting unit for the angle, because radian measurements are often treated as having either no dimensions or an implicitly assumed dimension. It is usually read as "−0.1952 radians per second," but it is only written as $\frac{-0.1952}{\text{sec}}$. It is not wrong, however, if you wrote $\frac{-0.1952 \, \text{radians}}{\text{sec}}$.

8. **Question:** A pebble is dropped into a pond, generating circular ripples. The radius of the largest ripple is increasing at a constant rate of 2 inches per second. After 5 seconds, what is the rate that the *circumference* of the largest ripple is increasing?

Solution: The rate that the circumference of the largest ripple is increasing is $12.5664 \, \frac{\text{in}}{\text{sec}}$

Explanation: This question explores the relationship between the radius and the circumference. This is given by the equation:

$$D = 2\pi R$$

To convert this into a related rate equation, we take the differential and then divide by dt:

$$D = 2\pi R$$

$$dD = 2\pi dR$$

$$\frac{dD}{dt} = 2\pi \frac{dR}{dt}$$

The problem states that the radius is increasing at a rate of $2 \, \frac{\text{in}}{\text{sec}}$, which is $\frac{dR}{dt}$. Note that this equation *doesn't have R* itself in the equation, so we don't need to solve for it, even though we could (therefore, the answer will be the same no

matter how many seconds we wait). Therefore, we can substitute this rate back into the related rate equation and get:

$$\frac{dD}{dt} = 2\pi \frac{dR}{dt}$$

$$\frac{dD}{dt} = 2\pi \, 2 \, \frac{\text{in}}{\text{sec}}$$

$$\frac{dD}{dt} = 4\pi \, \frac{\text{in}}{\text{sec}}$$

$$\frac{dD}{dt} \approx 12.5664 \, \frac{\text{in}}{\text{sec}}$$

9. **Question:** A hose is inflating a spherical baloon which begins empty. The water is coming out of the hose at a rate of $4 \, \frac{\text{mL}}{\text{sec}}$ (1mL = 1cm^3). After 10 seconds, what is the rate of the radius of the baloon increasing?

Solution: $0.0707 \, \frac{\text{cm}}{\text{sec}}$

Explanation: In solving this problem, we will use millimeters for length units to match the volume units.

So, the abstract geometry of the equation is that of a sphere.

The equation that relates the radius (r) to the volume (v) of a sphere is:

$$v = \frac{4}{3}\pi r^3$$

To convert this to a related rate equation, we simply take the differential and divide by dt:

$$v = \frac{4}{3}\pi r^3$$

$$dv = 3\frac{4}{3}\pi r^2 \, dr$$

$$dv = 4\pi r^2 \, dr$$

$$\frac{dv}{dt} = 4\pi r^2 \frac{dr}{dt}$$

We want to know the rate that the radius is increasing, which is $\frac{dr}{dt}$. So we can just solve for that:

$$\frac{dv}{dt} = 4\pi r^2 \frac{dr}{dt} \qquad \text{related rate equation}$$

$$\frac{1}{4\pi r^2} \frac{dv}{dt} = \frac{dr}{dt} \qquad \text{solved for } \frac{dr}{dt}$$

So, the pieces that we need to solve are the radius (r) and the rate of change in volume ($\frac{dr}{dt}$). The water coming through the hose is a change in *volume*. Therefore, the rate that the water coming through the hose is the rate that the volume of the sphere is increasing, and therefore represents $\frac{dv}{dt}$. This is given by the problem as $4\frac{\text{mL}}{\text{sec}}$.

The second thing we need is the radius. The radius starts at 0cm. However, we need to know what the radius will be *after* 10 *seconds*. Because the flow rate of the water is constant, we can easily calculate the *volume* after ten seconds:

$$4\frac{\text{mL}}{\text{sec}} \cdot 10\text{sec} = 40\text{mL}$$

We can then use the equation for the volume of a sphere to solve for r:

$$\frac{4}{3}\pi r^3 = v$$

$$\frac{4}{3}\pi r^3 = 40\text{mL}$$

$$r^3 = \frac{40\text{mL}}{\frac{4}{3}\pi}$$

$$r^3 = \frac{3 \cdot 40\text{mL}}{4\pi}$$

$$r^3 = \frac{30\text{mL}}{\pi}$$

$$r = \sqrt[3]{\frac{30\text{mL}}{\pi}}$$

$$r \approx 2.1216\text{cm}$$

Note in that last step, because $1\text{mL} = 1\text{cm}^3$, when we took the square root it switched from mL to cm because $\sqrt[3]{1\text{mL}} = \sqrt[3]{1\text{cm}^3} = 1\text{cm}$.

Now we have all of the pieces to substitute into our related rate equation:

$$\frac{dr}{dt} = \frac{1}{4\pi r^2}\frac{dv}{dt}$$

$$\frac{dr}{dt} = \frac{1}{4\pi(2.1216\text{cm})^2}(4\frac{\text{mL}}{\text{sec}})$$

$$\frac{dr}{dt} = \frac{1}{\pi(2.1216\text{cm})^2}\frac{\text{mL}}{\text{sec}}$$

$$\frac{dr}{dt} \approx \frac{1}{\pi 4.5012\text{cm}^2}\frac{\text{mL}}{\text{sec}}$$

$$\frac{dr}{dt} \approx \frac{1}{14.1410\text{cm}^2}\frac{\text{mL}}{\text{sec}}$$

$$\frac{dr}{dt} \approx 0.0707\frac{\text{mL}}{\text{cm}^2\,\text{sec}}$$

$$\frac{dr}{dt} \approx 0.0707\frac{\text{cm}^3}{\text{cm}^2\,\text{sec}}$$

$$\frac{dr}{dt} \approx 0.0707\frac{\text{cm}}{\text{sec}}$$

Therefore, after ten seconds, the radius of the baloon is increasing at a rate of $0.0707\frac{\text{cm}}{\text{sec}}$.

10. **Question:** A plane is flying at a constant altitude of 3 miles down a straight road at 150 miles per hour with a radar gun to catch speeding vehicles. The radar gun, however, only determines the speed of the vehicle *with relation to the airplane* (i.e., how fast the distance between the plane and the vehicle is increasing or decreasing) and the distance to the vehicle. If a vehicle headed *the same direction of the plane* is clocked by the radar gun as coming toward you at 80 miles per hour at a distance of 5 miles, how fast is the vehicle going?

Solution: $50\frac{\text{mi}}{\text{hr}}$

Explanation: If you think about the relationship between the plane, the ground, and the vehicle, you recognize that this is a right triangle. We will call the altitude of the plane A, the distance along the ground as G, and the direct distance through the air between the plane and the vehicle as D. Therefore, we know that the following relationship exists among these variables:

$$A^2 + G^2 = D^2$$

The altitude is given—3mi. The air distance is given by the radar gun—5mi. Therefore, we can easily enough figure out the ground distance, G:

$$A^2 + G^2 = D^2$$

$$G^2 = D^2 - A^2$$

$$G = \sqrt{D^2 - A^2}$$

$$G = \sqrt{(5)^2 - (3)^2}$$

$$G = 4\text{mi}$$

Now, we can convert the original equation into a differential equation and then to a related rate equation:

$$A^2 + G^2 = D^2$$

$$2A\,dA + 2G\,dG = 2D\,dD$$

$$A\,dA + G\,dG = D\,dD$$

$$A\frac{dA}{dt} + G\frac{dG}{dt} = D\frac{dD}{dt}$$

However, the plane is at a *constant* altitude. This means that $\frac{dA}{dt} = 0$. So we can simplify this to:

$$G\frac{dG}{dt} = D\frac{dD}{dt}$$

We are given $\frac{dD}{dt}$ from the radar gun: $-80\frac{\text{mi}}{\text{hr}}$. The value is negative because that is the speed is listed in the problem as "coming toward you." In other words, D is getting smaller, so $\frac{dD}{dt}$ must be negative. So we now have every variable we need to solve for $\frac{dG}{dt}$:

$$G\frac{dG}{dt} = D\frac{dD}{dt}$$

$$\frac{dG}{dt} = \frac{D}{G}\frac{dD}{dt}$$

$$\frac{dG}{dt} = \frac{5\text{mi}}{4\text{mi}} \cdot -80\frac{\text{mi}}{\text{hr}}$$

$$\frac{dG}{dt} = \frac{5}{4} \cdot -80\frac{\text{mi}}{\text{hr}}$$

$$\frac{dG}{dt} = -100\frac{\text{mi}}{\text{hr}}$$

It might be tempting to stop here. We know the ground speed by which the vehicle is closing on the plane. However, that is *not* the speed that the vehicle is going!

Our velocity closes the gap (i.e., makes G smaller), so it is counted as a negative contribution to $\frac{dG}{dt}$. However, the vehicle's velocity, since it is going the same direction, counterbalances that by *increasing* the gap (i.e.,

makes G bigger), so it is counted as a positive contribution to $\frac{dG}{dt}$. Therefore, to compute $\frac{dG}{dt}$, we will *add* the vehicles speed (which we will call $\frac{dV}{dt}$) and *subtract* the plane's speed (which we will call $\frac{dP}{dt}$, which is given in the problem as $150\frac{\text{mi}}{\text{hr}}$).

Therefore, we can think of the components of $\frac{dG}{dt}$ as follows:

$$\frac{dG}{dt} = \frac{dV}{dt} - \frac{dP}{dt}$$

The goal is to find $\frac{dV}{dt}$, so we can arrange this as:

$$\frac{dG}{dt} = \frac{dV}{dt} - \frac{dP}{dt}$$

$$\frac{dG}{dt} + \frac{dP}{dt} = \frac{dV}{dt}$$

$$-100\frac{\text{mi}}{\text{hr}} + 150\frac{\text{mi}}{\text{hr}} = \frac{dV}{dt}$$

$$\frac{dV}{dt} = 50\frac{\text{mi}}{\text{hr}}$$

So, the vehicle is traveling at $50\frac{\text{mi}}{\text{hr}}$.

11. **Question:** A radar gun manufacturer has asked you to develop an algorithm for the previous problem using only arithmetic operations, so that the operator only has to enter in the plane's speed and altitude, and it will calculate the speed of cars based on the measured distance and the measured relative speed. Develop a formula that will allow for this. You can assume that the car is headed in the same direction as the plane.

Solution: $\frac{dV}{dt} = \frac{D}{\sqrt{D^2 - A^2}}\frac{dD}{dt} + \frac{dP}{dt}$

Explanation: We can start this by using pieces from the previous question. The related rate equation we came up with was:

$$G\frac{dG}{dt} = D\frac{dD}{dt}$$

Likewise, we solved for G using the following formula:

$$G = \sqrt{D^2 - A^2}$$

We can put these together to come up with an equation solving for $\frac{dG}{dt}$ using only A, D, and $\frac{dD}{dt}$, which are all given in the equation:

$$G \frac{dG}{dt} = D \frac{dD}{dt}$$

$$\frac{dG}{dt} = \frac{D}{G} \frac{dD}{dt}$$

$$\frac{dG}{dt} = \frac{D}{\sqrt{D^2 - A^2}} \frac{dD}{dt}$$

Now, as we mentioned, we are not looking for $\frac{dG}{dt}$, but for the speed of the vehicle, $\frac{dV}{dt}$. The equation we came up with relating the plane's velocity to the vehicle's and the ground's is:

$$\frac{dG}{dt} = \frac{dV}{dt} - \frac{dP}{dt}$$

Therefore, we can substitute this into our original equation and solve for $\frac{dV}{dt}$:

$$\frac{dG}{dt} = \frac{D}{\sqrt{D^2 - A^2}} \frac{dD}{dt}$$

$$\frac{dV}{dt} - \frac{dP}{dt} = \frac{D}{\sqrt{D^2 - A^2}} \frac{dD}{dt}$$

$$\frac{dV}{dt} = \frac{D}{\sqrt{D^2 - A^2}} \frac{dD}{dt} + \frac{dP}{dt}$$

This is the value we are solving for, and all of the values on the right-hand side are given in the original equation. D is the radar-determined distance, $\frac{dD}{dt}$ is the radar-determined speed, A is the altitude, and $\frac{dP}{dt}$ is the plane's velocity.

Chapter 15

Working with Problems Involving Differentials

1. **Question:** Jan has a jug full of water. The amount of water in the jug is w. What term would you use to refer to the total rate of change in the amount of water in the jug?

   Solution: $\frac{dw}{dt}$

   Explanation: The differential is the result of all causes of change. Since we were asking for the total change in w, we can use the differential of w, dw. Since we are asking for a rate, we can divide it by dt.

2. **Question:** In the previous question, let's say that no water was being added, but water was leaving through (a) a leak, and (b) evaporation. Write an equation for the total rate of change of water that incorporates these causes.

   Solution: Either $\frac{dw}{dt} = \frac{de}{dt} + \frac{dl}{dt}$ or $\frac{dw}{dt} = -\frac{de}{dt} + -\frac{dl}{dt}$

   Explanation: In the previous question, we noted that the total rate of change is given by $\frac{dw}{dt}$. There are two ways that water is changing in this equation—by evaporation and by leaking. The most important thing is that evaporation and leaking be *different* terms than $\frac{dw}{dt}$, because $\frac{dw}{dt}$ only refers to the *total rate*. If we use e for the total amount of water that has evaporated, then $\frac{de}{dt}$ is the rate that evaporation is happening. If we use l for the amount of water that is leaking, then $\frac{dl}{dt}$ is the rate that leaking is happening. Therefore, $\frac{dw}{dt}$ is just the sum of these things:

$$\frac{dw}{dt} = -\frac{de}{dt} + -\frac{dl}{dt}$$

I used negative signs here, because the total water evaporated would be positive, but it would cause a removal of water from the jug. The same for leaking. It is not wrong to view these as positive as well, if you wanted them all normalized to the question of how much water is in the jug.

In that case, the equation would be:

$$\frac{dw}{dt} = \frac{de}{dt} + \frac{dl}{dt}$$

And the values of the rates on the right-hand side would be negative.

3. **Question:** The speed of gas flowing into a factory is twice the the speed of liquid leaving the factory. Write out this relationship using derivative notation.

   Solution: If we say that the amount of gas is g, the amount of liquid is l, and time is t, then we can say that $\frac{dg}{dt} = 2\frac{dl}{dt}$.

4. The bank account pays 4% interest per year (compounded continuously) on savings accounts. Write this out in differential notation.
   Solution: $\frac{dm}{dt} = 0.04m$

   Explanation: We can use the variable m to represent our current amount of money. Because the rate is 4% per year, this means that

$\frac{dm}{dt}$ is 0.04 times the amount of money in the bank. Therefore, the equation is $\frac{dm}{dt} = 0.04m$.

5. **Question:** If my position p along a road (in meters) after t seconds is given by the equation $p = t^2$, what is my velocity after 10 seconds.

Solution: $20\frac{\text{meter}}{\text{sec}}$

Explanation: Since velocity is the first derivative of position, we can find an equation for velocity by taking the derivative. if $p = t^2$, then $v = 2t$. Therefore, after 10 seconds, your velocity will be $20\frac{\text{meter}}{\text{sec}}$.

6. **Question:** The number of students at the school is given by S. The number of students change throughout the year due to enrollment (which adds students), graduation (which removes them), and drop-outs (which also removes them). Write an equation for the change in students over time.

Solution: Depending on how the sign of the quantities are treated, the equation can be either:
$$\frac{dS}{dt} = \frac{de}{dt} - \frac{dg}{dt} - \frac{dd}{dt}$$
or
$$\frac{dS}{dt} = \frac{de}{dt} + \frac{dg}{dt} + \frac{dd}{dt}$$

Explanation: Because the number of students is S, the change in students is dS. Now, S is technically discrete, but, if we have a large enough student population, we can treat it as continuous. Therefore, the change in student population over time is $\frac{dS}{dt}$. The enrollment rate can be listed as $\frac{de}{dt}$. The graduation rate can be listed as $\frac{dg}{dt}$. The drop-out rate can be listed as $\frac{dd}{dt}$. The change in student population will be a combination of all of these:
$$\frac{dS}{dt} = \frac{de}{dt} - \frac{dg}{dt} - \frac{dd}{dt}$$

Alternatively, if you normalized everything according to its relationship to S, then you would have:
$$\frac{dS}{dt} = \frac{de}{dt} + \frac{dg}{dt} + \frac{dd}{dt}$$

In this case, the graduation rate ($\frac{dg}{dt}$) and the drop-out rate ($\frac{dd}{dt}$) would be listed as negative quantities.

7. **Question:** If my position p along a road (in meters) after t seconds is given by the equation $p = t^3$, what is my velocity after 10 seconds?

Solution: $300\frac{\text{meters}}{\text{second}}$

Explanation: Velocity is the first derivative of position. Therefore, if $p = t^3$, $v = 3t^2$. At $t = 10$, this will be $3(10)^2$, or 300. Since we are dealing in meters and seconds, this is $300\frac{\text{meters}}{\text{second}}$.

8. **Question:** If my position p along a road (in meters) after t seconds is given by the equation $p = 5t$, what is my acceleration?

Solution: The acceleration is zero.

Explanation: Acceleration is the second derivative of position. If $p = 5t$, the first derivative is $v = 5$. The second derivative is $a = 0$.

9. **Question:** In a given engine, there is usually an optimal speed at which the engine runs (given as revolutions-per-minute, or RPM), such that its ability to convert energy into work is maximized. If I had an equation which related RPM to the ability to convert energy to work, how would I (in general terms) find the maximum value?

Solution: In an equation relating two values, the maximum can be found by taking the derivative and finding where the derivative is zero. Here, the independent variable is RPM (we will call this r), and the value is the energy-to-work conversion ratio (we will call this w). Therefore, to find the maximum value, we look for the location where the slope between these is zero. In other words, we solve for r where $\frac{dw}{dr} = 0$.

Part III

The Integral

Chapter 16

The Integral as an Antidifferential

Integrate the following:

1. **Question:** $\int dx$

   Solution: $x + C$

   Explanation: Simple variable rule

2. **Question:** $\int x \, dx$

   Solution: $\frac{x^2}{2} + C$

   Explanation: Power rule

3. **Question:** $\int x^7 \, dx$

   Solution: $\frac{x^8}{8} + C$

   Explanation: Power rule

4. **Question:** $\int (x \, dx + x^2 \, dx)$

   Solution: $\frac{x^2}{2} + \frac{x^3}{3} + C$

   Explanation: Power and addition rules

5. **Question:** $\int 3e^x \, dx$

   Solution: $3e^x + C$

   Explanation: Constant multiplier rule and exponent rule

6. **Question:** $\int \frac{1}{x} \, dx$

   Solution: $\ln(x) + C$

   Explanation: Logarithm rule

7. **Question:** $dy = 3x^2 \, dx$

   Solution: $y = x^3 + C$

   Explanation:
   $$dy = 3x^2 \, dx$$
   $$\int dy = \int 3x^2 \, dx$$
   $$\int dy = 3 \int x^2 \, dx$$
   $$y = 3\frac{x^3}{3} + C$$
   $$y = x^3 + C$$

8. **Question:** $y \, dy = \sin(x) \, dx$

   Solution: $\frac{y^2}{2} = -\cos(x) + C$

   Explanation:
   $$y \, dy = \sin(x) \, dx$$
   $$\int y \, dy = \int \sin(x) \, dx$$
   $$\frac{y^2}{2} = -\cos(x) + C$$

Explanation:

$$\cos(y)\,dy - y^2\,dy = \sin(x)\,dx + 5x^2\,dx$$

$$\int (\cos(y)\,dy - y^2\,dy) = \int (\sin(x)\,dx + 5x^2\,dx)$$

$$\int \cos(y)\,dy - \int (y^2\,dy) = \int \sin(x)\,dx + \int 5x^2\,dx$$

$$\sin(y) - \frac{y^3}{3} = -\cos(x) + 5\frac{x^3}{3} + C$$

9. **Question:** $\frac{dy}{dx} = \cos(x) + 2^x$

   Solution: $y = \sin(x) + \frac{2^x}{\ln(2)} + C$

   Explanation:

   $$\frac{dy}{dx} = \cos(x) + 2^x$$

   $$dy = \cos(x)\,dx + 2^x\,dx$$

   $$\int dy = \int (\cos(x)\,dx + 2^x\,dx)$$

   $$\int dy = \int \cos(x)\,dx + \int 2^x\,dx$$

   $$y = \sin(x) + \frac{2^x}{\ln(2)} + C$$

10. **Question:** $\frac{dy}{dx} = \tan(x) - e^x - x^4$

    Solution: $y = -\ln(|\cos(x)|) - e^x - \frac{x^5}{5} + C$

    Explanation:

    $$\frac{dy}{dx} = \tan(x) - e^x - x^4$$

    $$dy = \tan(x)\,dx - e^x\,dx - x^4\,dx$$

    $$\int dy = \int (\tan(x)\,dx - e^x\,dx - x^4\,dx)$$

    $$\int dy = \int \tan(x)\,dx - \int e^x\,dx - \int x^4\,dx$$

    $$y = -\ln(|\cos(x)|) - e^x - \frac{x^5}{5} + C$$

11. **Question:** $\cos(y)\,dy - y^2\,dy = \sin(x)\,dx + 5x^2\,dx$

    Solution: $\sin(y) - \frac{y^3}{3} = -\cos(x) + 5\frac{x^3}{3} + C$

Chapter 17

The Integral as an Infinite Sum

Find the following definite integrals:

1. **Question:** $\int_1^2 x \, dx$

   Solution: $\frac{3}{2}$

   Explanation: To find the definite integral, we first convert this to an indefinite integral with an evaluation bar. Then, we solve for the integral. Finally, we evaluate the new function using the evaluation bar.

   $$\int_1^2 x \, dx = \int x \, dx \Big|_1^2$$
   $$= \frac{x^2}{2} \Big|_1^2$$
   $$= \frac{(2)^2}{2} - \frac{(1)^2}{2}$$
   $$= 2 - \frac{1}{2}$$
   $$= \frac{3}{2}$$

2. **Question:** $\int_7^{10} 6x^2 \, dx$

   Solution: 1314

   Explanation: To find the definite integral, we first convert this to an indefinite integral with an evaluation bar. Then, we solve for the

integral and evaluate:

$$\int_7^{10} 6x^2 \, dx = \int 6x^2 \, dx \Big|_7^{10}$$
$$= \frac{6x^3}{3} \Big|_7^{10}$$
$$= 2x^3 \Big|_7^{10}$$
$$= 2(10)^3 - 2(7)^3$$
$$= 1314$$

3. **Question:** $\int_{-6}^{-2} (3x^2 - 4x + 3) \, dx$

   Solution: 284

   Explanation: Again, to find the definite integral, we first convert this to an indefinite integral with an evaluation bar. Then, we solve for the integral. Finally, we evaluate the new function using the evaluation bar.

   $$\int_{-6}^{-2} (3x^2 - 4x + 3) \, dx$$
   $$= \int (3x^2 - 4x + 3) \, dx \Big|_{-6}^{-2}$$
   $$= \int 3x^2 \, dx - 4x \, dx + 3 \, dx \Big|_{-6}^{-2}$$
   $$= x^3 - 2x^2 + 3x \Big|_{-6}^{-2}$$
   $$= (-2)^3 - 2(-2)^2 + 3(-2) - ((-6)^3 - 2(-6)^2 + 3(-6))$$
   $$= -8 - 8 + -6 - (-216 - 72 + -18)$$
   $$= 284$$

Explanation:

$$\int_{-1}^{4} e^x \, dx = \int e^x \, dx \Big|_{-1}^{4}$$

$$= e^x \Big|_{-1}^{4}$$

$$= e^4 - e^{-1}$$

$$\approx 54.5982 - 0.3679$$

$$\approx 54.2303$$

4. **Question:** $\int_0^1 e^x \, dx$

Solution: $e - 1$ or 1.7183

Explanation:

$$\int_0^1 e^x \, dx = \int e^x \, dx \Big|_0^1$$

$$= e^x \Big|_0^1$$

$$= e^1 - e^0$$

$$= e - 1$$

$$\approx 1.7183$$

7. **Question:** $\int_{x=2,z=3}^{x=5,z=6} x^2 \, dx - 2^z \, dz$

Solution: -75.1242

Explanation: This is just like the previous problems, but the limits of integration are more complicated because we are dealing in multiple variables.

$$\int_{x=2,z=3}^{x=5,z=6} x^2 \, dx - 2^z \, dz = \int (x^2 \, dx - 2^z \, dz) \Big|_{x=2,z=3}^{x=5,z=6}$$

$$= \int x^2 \, dx - \int 2^z \, dz \Big|_{x=2,z=3}^{x=5,z=6}$$

$$= \frac{x^3}{3} - \frac{2^z}{\ln(2)} \Big|_{x=2,z=3}^{x=5,z=6}$$

$$= \frac{(5)^3}{3} - \frac{2^6}{\ln(2)} - \left(\frac{(2)^3}{3} - \frac{2^3}{\ln(2)}\right)$$

$$= \frac{125}{3} - \frac{64}{\ln(2)} - \left(\frac{8}{3} - \frac{8}{\ln(2)}\right)$$

$$\approx 41.6667 - 92.3324 - (2.6667 - 11.541)$$

$$\approx -41.7908$$

5. **Question:** $\int_0^{2\pi} \cos(x) \, dx$

Solution: 0

Explanation:

$$\int_0^{2\pi} \cos(x) \, dx = \int \cos(x) \, dx \Big|_0^{2\pi}$$

$$= \sin(x) \Big|_0^{2\pi}$$

$$= \sin(2\pi) - \sin(0)$$

$$= 0 - 0$$

$$= 0$$

Find the given infinite sums:

6. **Question:** $\int_{-1}^4 e^x \, dx$

Solution: 54.2303

8. **Question:** Find the infinite sum of every value of $5x \, dx$ from $x = 2$ to $x = 8$.

Solution: 150

Explanation: Finding an infinite sum of infinitesimals is the same thing as finding the definite integral.

$$\int_2^8 5x\,dx = \int 5x\,dx \Big|_2^8$$

$$= \frac{5}{2}x^2 \Big|_2^8$$

$$= \frac{5}{2}(8)^2 - \frac{5}{2}(2)^2$$

$$= 160 - 10$$

$$= 150$$

The infinite sum of all infinitesimal values $5x\,dx$ from 2 to 8 is 150.

9. **Question:** Find the infinite sum of every dy from $x = -3$ to $x = 4$ where $dy = \sin(x)\,dx + x\,dx$

Solution: 3.1636

Explanation: Since this is an equation, we just need to isolate dy by itself and integrate both sides. In this equation, dy is already isolated, so we just need to integrate both sides:

$$dy = \sin(x)\,dx + x\,dx$$

$$\int_{-3}^4 dy = \int_{-3}^4 (\sin(x)\,dx + x\,dx)$$

$$= \int (\sin(x)\,dx + x\,dx) \Big|_{-3}^4$$

$$= \int \sin(x)\,dx + \int x\,dx \Big|_{-3}^4$$

$$= -\cos(x) + \frac{x^2}{2} \Big|_{-3}^4$$

$$= -\cos(4) + \frac{(4)^2}{2} - (-\cos(-3) + \frac{(-3)^2}{2})$$

$$= -(-0.6536) + 8 - (-(-0.9900) + 4.5)$$

$$= 3.1636$$

10. **Question:** Find the infinite sum of every dy from $x = 1$ to $x = 2$, where $\frac{dy}{dx} = \frac{9}{x}$.

Solution: 6.2383

Explanation: In order to find the infinite sum of all dy, we have to isolate dy to one side

of the equation. We can do this by multiplying both sides by dx, and then proceeding as before:

$$\frac{dy}{dx} = \frac{9}{x}$$

$$dy = \frac{9}{x}\,dx$$

$$\int_1^2 dy = \int_1^2 \frac{9}{x}\,dx$$

$$= \int \frac{9}{x}\,dx \Big|_1^2$$

$$= 9\ln(x) \Big|_1^2$$

$$= 9\ln(2) - 9\ln(1)$$

$$= 6.2383$$

The sum of all values of dy from 1 to 2 is 6.2383.

11. **Question:** Find the sum of every dx from $y = 2, z = 3$ to $y = 6, z = 8$ for the equation $dx = y\,dy - 2z\,dz$

Solution: -39

Explanation: Because the equation is already solved for dx, we can simply integrate the equation:

$$dx = y\,dy - 2z\,dz$$

$$\int_{y=2,z=3}^{y=6,z=8} dx = \int_{y=2,z=3}^{y=6,z=8} y\,dy - 2z\,dz$$

$$= \int y\,dy - 2z\,dz \Big|_{y=2,z=3}^{y=6,z=8}$$

$$= \frac{y^2}{2} - z^2 \Big|_{y=2,z=3}^{y=6,z=8}$$

$$= \left(\frac{(6)^2}{2} - (8)^2\right) - \left(\frac{(2)^2}{2} - (3)^2\right)$$

$$= (18 - 64) - (2 - 9)$$

$$= -39$$

12. **Question:** Find the sum of every dq from $x = 2, y = 1$ to $x = -1, y = 3$ for the equation $\frac{dq}{dx} + \frac{dy}{dx} = 5$

Solution: −17

Explanation: To solve this equation, first we have to solve for dq:

$$\frac{dq}{dx} + \frac{dy}{dx} = 5$$

$$dq + dy = 5\,dx$$

$$dq = 5\,dx - dy$$

Now, to find the infinite sum of dq over these values, we simply integrate both sides:

$$dq = 5\,dx - dy$$

$$\int_{x=2,y=1}^{x=-1,y=3} dq = \int_{x=2,y=1}^{x=-1,y=3} 5\,dx - dy$$

$$= \int 5\,dx - dy \Big|_{x=2,y=1}^{x=-1,y=3}$$

$$= 5x - y \Big|_{x=2,y=1}^{x=-1,y=3}$$

$$= (5(-1) - (3)) - (5(2) - (1))$$

$$= -5 - 3 - 10 + 1$$

$$= -17$$

Chapter 18

Using the Integral to Find the Area Under the Curve

1. **Question:** Find the area under the curve $y = 4$ from $x = 0$ to $x = 3$ using the method specified in this chapter. Then, draw the equation and see if there is some other way that you know of obtaining the area.

    Solution: The area under the curve is 12 square units. Since the shape formed by the line is merely a rectangle, the area could have also been found by multiplying width times height.

    Explanation: The formula for the area under the curve is given in Equation 18.1:

    $$\text{area} = \int_{x=\text{start}}^{x=\text{finish}} y \, dx$$

    In the question y is given as just 4, the start is 0, and the finish is 3. This yields the new equation:

    $$\text{area} = \int_{x=0}^{x=3} 4 \, dx$$

    This can be easily solved as a definite integral:

    $$\text{area} = \int_{x=0}^{x=3} 4 \, dx$$
    $$= \int 4 \, dx \Big|_{x=0}^{x=3}$$
    $$= 4x \Big|_{x=0}^{x=3}$$
    $$= 4(3) - 4(0) \qquad = 12$$

    Therefore, the area is 12 square units.

    For the second part of the question, if we were to draw the equation, it is just a straight horizontal line where $y = 4$. If you shade in the part between $x = 0$ and $x = 3$ it becomes a normal rectangle.

Rectangles are easy to find the area of—just width times height. The height is 4, and the width is just the distance from the start to the finish ($3 - 0 = 3$). Therefore, we can calculate the area as:

$$\text{area} = \text{width} \cdot \text{height}$$
$$= 3 \cdot 4$$
$$= 12$$

This agrees with the other way of calculating area.

2. **Question:** Find the area under the curve $y = 4x - 2$ from $x = 2$ to $x = 4$. What shape does this form on the graph?

    Solution: 20 square units. This graph represents a triangle.

    Explanation: The formula for the area under the curve is given in Equation 18.1:

    $$\text{area} = \int_{x=\text{start}}^{x=\text{finish}} y \, dx$$

    We can then plug in the formula for y (as well as the other starting values) into this equation to yield a definite integral:

    $$\text{area} = \int_{x=2}^{x=4} (4x - 2) \, dx$$

We can now solve for the area directly:

$$\text{area} = \int_{x=2}^{x=4} (4x - 2)\,dx$$

$$= \int 4x\,dx - 2\,dx \Big|_2^4$$

$$= \frac{4x^2}{2} - 2x \Big|_2^4$$

$$= \frac{4(4)^2}{2} - 2(4) - \left(\frac{4(2)^2}{2} - 2(2) \right)$$

$$= 32 - 8 - 8 + 4$$

$$= 20$$

3. **Question:**   Find the area under the curve $y = x^2 - 2x$ from $x = -1$ to $x = 0$

Solution: $\frac{4}{3}$

Explanation: To find the area under a given curve, first we need to use Equation 18.1 to convert this into an integral:

$$\text{area} = \int_{start}^{finish} y\,dx$$

$$= \int_{-1}^{0} x^2\,dx - 2x\,dx$$

$$= \int x^2\,dx - 2x\,dx \Big|_{-1}^0$$

$$= \frac{x^3}{3} - x^2 \Big|_{-1}^0$$

$$= \frac{(0)^3}{3} - (0)^2 - (\frac{(-1)^3}{3} - (-1)^2)$$

$$= 0 + \frac{1}{3} + 1$$

$$= \frac{4}{3}$$

4. **Question:**   Find the area under the curve $y = e^x - \sin(x)$ from $x = 0$ to $x = \pi$.

Solution: $e^\pi - 3$ or 20.1407

Explanation: To find the area under a

given curve, first we need to use Equation 18.1 to convert this into an integral:

$$\text{area} = \int_{start}^{finish} y\,dx$$

$$= \int_0^\pi (e^x - \sin(x))\,dx$$

$$= \int (e^x - \sin(x))\,dx \Big|_0^\pi$$

$$= \int e^x\,dx - \sin(x)\,dx \Big|_0^\pi$$

$$= e^x + \cos(x) \Big|_0^\pi$$

$$= e^\pi + \cos(\pi) - (e^0 + \cos(0))$$

$$= e^\pi + -1 - (1 + 1)$$

$$= e^\pi - 3$$

$$\approx 20.1407$$

5. **Question:** Find the area under the curve $y = \frac{1}{x}$ from $x = 2$ to $x = 3$

Solution: 0.4055 square units

Explanation: To find the area under a given curve, first we need to use Equation 18.1 to convert this into an integral:

$$\text{area} = \int_{start}^{finish} y\,dx$$

$$= \int_2^3 \frac{1}{x}\,dx$$

$$= \int \frac{1}{x}\,dx \Big|_2^3$$

$$= \ln(x) \Big|_2^3$$

$$= \ln(3) - \ln(2)$$

$$\approx 1.0986 - 0.6931$$

$$\approx 0.4055$$

6. **Question:**   Find the area under the curve $y = \frac{1}{x^2}$ from $x = 1$ to $x = 5$. You may need to

think about other ways to write this in order to solve it.

Solution: The area under the curve is $\frac{4}{5}$ square units.

Explanation: To find the area under a given curve, first we need to use Equation 18.1 to convert this into an integral:

$$\text{area} = \int_{start}^{finish} y \, dx$$

$$= \int_1^5 \frac{1}{x^2} \, dx$$

You may think that you don't know how to integrate this. Or, worse, you may think that it looks like it integrates to some form of natural log. Neither are true. You can simply convert it to a negative exponent and then use the power rule.

$$\text{area} = \int_1^5 \frac{1}{x^2} \, dx$$

$$= \int_1^5 x^{-2} \, dx$$

$$= \int x^{-2} \, dx \Big|_1^5$$

$$= \frac{x^{-1}}{-1} \Big|_1^5$$

$$= -\frac{1}{x} \Big|_1^5$$

$$= -\frac{1}{5} - -\frac{1}{1}$$

$$= -\frac{1}{5} + 1$$

$$= \frac{4}{5}$$

7. **Question:** Find the area under the curve $y = \ln(x)$ from $x = 1$ to $x = 7$. Note that this integral is not one that you would already know from differentiation, so you need to look it up in Appendix H.7.

Solution: 7.6213 square units

Explanation: To find the area under a given curve, first we need to use Equation 18.1

to convert this into an integral:

$$\text{area} = \int_{start}^{finish} y \, dx$$

$$= \int_1^7 \ln(x) \, dx$$

$$= \int \ln(x) \, dx \Big|_1^7$$

$$= x \ln(x) - x \Big|_1^7$$

$$= 7 \ln(7) - 7 - (1 \ln(1) - 1)$$

$$\approx 7 \cdot 1.9459 - 7 - (0 - 1)$$

$$\approx 7.6213$$

8. **Question:** Find the area under the curve $y = -x^2 - 2x - 1$ from $x = 2$ to $x = 4$.

Solution: $-\frac{98}{3}$ square units

Explanation: To find the area under a given curve, first we need to use Equation 18.1 to convert this into an integral:

$$\text{area} = \int_{start}^{finish} y \, dx$$

$$= \int_2^4 (-x^2 - 2x - 1) \, dx$$

$$= \int -x^2 \, dx - 2x \, dx - dx \Big|_2^4$$

$$= -\frac{x^3}{3} - x^2 - x \Big|_2^4$$

$$= -\frac{(4)^3}{3} - (4)^2 - (4) - (-\frac{(2)^3}{3} - (2)^2 - 2)$$

$$= -\frac{64}{3} - 16 - 4 + \frac{8}{3} + 4 + 2$$

$$= -\frac{56}{3} - 14$$

$$= -\frac{98}{3}$$

You may find it strange to get a negative area. However, we are looking for the area *under the curve*. If you graph this, you will find that the curve is actually *below* the x-axis. That means that the area "under" the curve is actually above it (i.e., between the curve and the x-axis).

Therefore, the negative value just means that the curve has more area below the x-axis than above it.

the area under a curve are inverses.

9. **Question:** Find the area *between* the curves $y = \sin(x)$ and $y = -x^2$ from $x = 0$ to $x = \pi$.

Solution: 12.3354 square units

Explanation: Finding the area between two curves can be done just by the finding the area beneath each curve and subtracting. $y = \sin(x)$ is the top curve, so we will start there:

$$\text{area} = \int_0^\pi \sin(x)\,dx$$

$$= \int \sin(x)\,dx \,\Big|_0^\pi$$

$$= -\cos(x) \,\Big|_0^\pi$$

$$= -\cos(\pi) - -\cos(0)$$

$$= -(-1) + 1$$

$$= 2$$

Now we will find the area beneath $y = -x^2$:

$$\text{area} = \int_0^\pi -x^2\,dx$$

$$= \int -x^2\,dx \,\Big|_0^\pi$$

$$= -\frac{x^3}{3} \,\Big|_0^\pi$$

$$= -\frac{(\pi)^3}{3} - \frac{(0)^3}{3}$$

$$\approx -10.3354$$

The total will be the top area (2) minus the bottom area (−10.3354), which yields 12.3354 square units.

10. **Question:** Which of the following operations are inverses of each other: differentiation, finding a derivative, integration, finding the area under a curve.

Solution: Differentiation and integration are inverses. Finding a derivative and finding

Chapter 19

Integration and u-Substitution

Find the indefinite integrals:

1. **Question:** $\int (1+x)^2 \, dx$

   Solution: $\frac{(1+x)^3}{3} + C$

   Explanation: In this case, notice that x differs from $x+1$ by only a constant. This is a great candidate for u-substitution because the differential for both of them will be the same. We will set $u = 1+x$ and therefore $du = dx$. This problem now becomes:

   $$\int u^2 \, du$$

   This is a straightforward application of the power rule:

   $$\int u^2 \, du = \frac{u^3}{3} + C$$

   Now we can simply replace u with $1+x$, which yields $\frac{(1+x)^3}{3} + C$.

2. **Question:** $\int \cos(2x) \, dx$

   Solution: $\frac{1}{2}\sin(2x) + C$

   Explanation: In this case, the differential of $2x$ is very close to just the differential of x, so it makes a great candidate for u-substitution. We will set $u = 2x$, which means that $du = 2\,dx$. However, the equation only has dx, not $2\,dx$.

   Therefore, dividing both sides by 2 gives us $\frac{du}{2} = dx$. Now our equation becomes:

   $$\int \cos(u) \, \frac{du}{2}$$

We can move the division by two out in front as a constant multiplier, which yields:

$$\frac{1}{2} \int \cos(u) \, du$$

This integral is easy to solve! It yields:

$$\frac{1}{2} \sin(u) + C$$

Now we just substitute back in $2x$ for u, givin us:

$$\frac{1}{2} \sin(2x) + C$$

3. **Question:** $\int \cos(x^2) 2x \, dx$

   Solution: $\sin(x^2) + C$

   Explanation: Always remember when integrating that if there is a function of something, look outside of the function to see if something *resembling* its differential is there. In this case, the inside of the cosine function is x^2. What is the differential of x^2? It is $2x \, dx$, which is exactly what we have on the outside of our function as well.

   Therefore, we can set $u = x^2$, which means that $du = 2x \, dx$. This makes a straightforward replacement into our integral, yielding:

   $$\int \cos(u) \, du$$

   This yields:

   $$\sin(u) + C$$

   Then, substituting back in for u gives us the final answer:

   $$\sin(x^2) + C$$

Moving the constant out in front gives:

$$\frac{1}{2} \int u^{\frac{1}{2}} \, du$$

Applying the power rule we get:

$$\frac{1}{2} \frac{u^{\frac{3}{2}}}{\frac{3}{2}} + C$$

This simplifies to:

$$\frac{1}{3} u^{\frac{3}{2}} + C$$

Substituting back in for u gives us:

$$\frac{1}{3}(2x - 1)^{\frac{3}{2}}$$

4. **Question:** $\int \sin(x^3) x^2 \, dx$

Solution: $-\frac{1}{3} \cos(x^3) + C$

Explanation: This is very similar to the previous problem, except that $x^2 \, dx$ is not *exactly* the differential of x^3, so we will have to massage it a little bit. But, to start out with, let's set $u = x^3$. That means that $du = 3x^2 \, dx$. We need it to match $x^2 \, dx$. To do that, we just divide both sides by three, giving us the equation $\frac{du}{3} = x^2 \, dx$.

Now we substitute that into the integral, giving us:

$$\int \sin(u) \frac{du}{3}$$

The $\frac{1}{3}$ can be moved out in front as a constant multiplier, leaving us with:

$$\frac{1}{3} \int \sin(u) \, du$$

This gives us:

$$-\frac{1}{3} \cos(u) + C$$

Substituting back in for u gives us:

$$-\frac{1}{3} \cos(x^3) + C$$

6. **Question:** $\int \frac{e^{x^2} x}{3} \, dx$

Solution: $\frac{1}{6} e^{2x} + C$

Explanation: In this problem, we should start by moving the division by three outside of the integral to make like easier for us:

$$\frac{1}{3} \int e^{x^2} x \, dx$$

Now, $x \, dx$ is almost the differential of x^2, so let's set $u = x^2$ and see what we can do. If $u = x^2$, then $du = 2x \, dx$, but we just want $x \, dx$, so we will divide both sides by two, giving $\frac{du}{2} = x \, dx$. This yields:

$$\frac{1}{3} \int e^u \frac{du}{2}$$

Moving the $\frac{1}{2}$ out front gives us:

$$\frac{1}{6} \int e^u \, du$$

The integral is simply:

$$\frac{1}{6} e^u + C$$

Substituting back in for u gives us:

$$\frac{1}{6} e^{2x} + C$$

5. **Question:** $\int \sqrt{2x - 1} \, dx$

Solution: $\frac{1}{3}(2x - 1)^{\frac{3}{2}}$

Explanation: To do this one, remember that a square root can be converted into an exponent. This then becomes:

$$\int (2x - 1)^{\frac{1}{2}} \, dx$$

If we set $u = 2x - 1$, then $du = 2 \, dx$. But we just want dx by itself. Therefore, we can rearrange to say $\frac{du}{2} = dx$.

The problem now becomes:

$$\int u^{\frac{1}{2}} \frac{du}{2}$$

Find the definite integral:

7. **Question:** $\int_2^7 (x - 34)^2 \, dx$

   Solution: 4361.6667

   Explanation: To solve this, we recognize that:

   $$\int_2^7 (x - 34)^2 \, dx = \int (x - 34)^2 \, dx \Big|_2^7$$

   So, first we need to evaluate the integral. Using u-substitution, we can say that $u = x - 34$, so $du = dx$. This becomes $\int u^2 \, du$. Using the power rule, this integrates to $\frac{u^3}{3} + C$. If we substitute back in for u, that gives us $\frac{(x-34)^3}{3} + C$.

   Now we need to evaluate it at 2 and 7. This gives:

   $$\frac{(7 - 34)^3}{3} - \frac{(2 - 34)^3}{3} \approx 4361.6667$$

8. **Question:** $\int_3^{3.25} \sin(x^4 + 6) x^3 \, dx$

   Solution: 0.2027

   Explanation: First, let's convert this to indefinite integration plus evaluation:

   $$\int_3^{3.25} \sin(x^4 + 6) x^3 \, dx = \int \sin(x^4 + 6) x^3 \, dx \Big|_3^{3.25}$$

   If we set $u = x^4 + 6$, then $du = 4x^3 \, dx$. But we need $x^3 \, dx$, so we divide both sides by four. This gives $\frac{du}{4} = x^3 \, dx$. Now we can make our substitutions, which give:

   $$\int \sin(u) \frac{du}{4}$$

   Now we can move the constant divisor outside of the integral:

   $$\frac{1}{4} \int \sin(u) \, du$$

   This integrates to:

   $$-\frac{1}{4} \cos(u) + C$$

   Substituting back in for u gives us:

   $$-\frac{1}{4} \cos(x^4 + 6) + C$$

Now we need to apply the evaluation bar:

$$\int_3^{3.25} \sin(x^4 + 6) x^3 \, dx$$

$$= -\frac{1}{4} \cos((3.25)^4 + 6) - -\frac{1}{4} \cos((3)^4 + 6)$$

$$= -\frac{1}{4} \cos((3.25)^4 + 6) + \frac{1}{4} \cos((3)^4 + 6)$$

$$= -\frac{1}{4}(-0.2409) + \frac{1}{4}(0.5698)$$

$$= 0.2027$$

Integrate the following equations:

9. **Question:** $\frac{dy}{dx} = (x - 3)^5$

   Solution: $y = \frac{(x-3)^6}{6} + C$

   Explanation: To solve this, first we need to convert this into an equation that can be integrated. In other words, we need to multiply both sides by dx. This gives:

   $$dy = (x - 3)^5 \, dx$$

   Using u-substitution, we can set $u = x - 3$, and $du = dx$. This give:

   $$dy = u^5 \, du$$

   Integrating both sides gives us:

   $$y = \frac{u^6}{6} + C$$

   Substituting back in for u gives us:

   $$y = \frac{(x - 3)^6}{6} + C$$

10. **Question:** $e^{2y} \, dy = \cos(x - 5) \, dx$

    Solution: $\frac{1}{2} e^{2y} = \sin(x - 5) + C$

    Explanation: To solve this integral, you need to simply integrate both sides of the equation. This yields:

    $$\int e^{2y} \, dy = \int \cos(x - 5) \, dx$$

On the left-hand side, we can do a u-substitution of $u = 2y$ and therefore $du = 2dy$. We need just dy, so we can rearrange this into $\frac{du}{2} = dy$. Therefore, the left-hand side gives us:

$$\int e^u \frac{du}{2}$$

This simplifies to:

$$\frac{1}{2} \int e^u \, du$$

This integrates to:

$$\frac{1}{2} e^u + C$$

We can substitute back in for u, giving us:

$$\frac{1}{2} e^{2y} + C$$

On the right-hand side, we can say that $u = x - 5$, and $du = dx$. The right-hand side then becomes:

$$\int \cos(u) \, du$$

This integrates to:

$$\sin(u) + C$$

Substituting back in for u gives us:

$$\sin(x - 5) + C$$

Putting together both sides we get:

$$\frac{1}{2} e^{2y} + C_1 = \sin(x - 5) + C_2$$

However, we only need one constant of integration, so we just pick one and get:

$$\frac{1}{2} e^{2y} = \sin(x - 5) + C$$

Find the area under each curve:

11. **Question:** $y = (x - 10)^7$ from $x = 12$ to $x = 14$

    Solution: The area is 8160 square units.

    Explanation: To find the area under a

given curve, first we need to use Equation 18.1 to convert this into an integral:

$$\text{area} = \int_{start}^{finish} y \, dx$$

$$= \int_{12}^{14} (x - 10)^7 \, dx$$

$$= \int (x - 10)^7 \, dx \Big|_{12}^{14}$$

The next step is to solve the integral $\int (x - 10)^7 \, dx$. While we *could* expand this out into a giant polynomial, doing u-substitution is much easier. We will set $u = x - 10$. Differentiating both sides of this we find $du = dx$.

Now our integral is $\int u^7 \, du$. This becomes $\frac{u^8}{8}$. Backsubstituting in for u gives us $\frac{(x-10)^8}{8}$. Now we just need to evaluate this at the limits of integration:

$$\text{area} = \frac{(x - 10)^8}{8} \Big|_{12}^{14}$$

$$= \frac{((14) - 10)^8}{8} - \frac{((12) - 10)^8}{8}$$

$$= \frac{4^8}{8} - \frac{2^8}{8}$$

$$= 8192 - 32$$

$$= 8160$$

12. **Question:** $y = \sin(x^3) \, x^2 \, dx$ from $x = 0$ to $x = 1$.

    Solution: Approximately 0.1532 square units.

    Explanation: To find the area under a given curve, first we need to use Equation 18.1 to convert this into an integral:

$$\text{area} = \int_{start}^{finish} y \, dx$$

$$= \int_{0}^{1} \sin(x^3) \, x^2 \, dx$$

$$= \int \sin(x^3) \, x^2 \, dx \Big|_{0}^{1}$$

To solve $\int \sin(x^3) \, x^2 \, dx$ we need to do a u-substitution. We will start by setting $u = x^3$.

Next, we will find du by differentiation: $du = 3x^2 \, dx$. However, to solve the integral, we have to be able to replace $x^2 \, dx$. If we divide both sides by 3 we will get $\frac{du}{3} = x^2 \, dx$. This can be used as a replacement.

Now we have:

$$\text{area} = \int \sin(u) \frac{du}{3} \Big|_{x=0}^{x=1}$$

$$= \frac{1}{3} \int \sin(u) \, du \Big|_{x=0}^{x=1} \qquad = -\frac{1}{3} \cos(u) \, du \Big|_{x=0}^{x=1}$$

$$= -\frac{1}{3} \cos(x^3) \, du \Big|_{x=0}^{x=1}$$

$$= -\frac{1}{3} \cos((1)^3) - -\frac{1}{3} \cos((0)^3)$$

$$= -\frac{1}{3} \cos(1) + \frac{1}{3} \cos(0)$$

$$\approx -\frac{1}{3}(0.5403) + \frac{1}{3}(1)$$

$$\approx -0.1801 + 0.3333$$

$$\approx 0.1532$$

Chapter 20

Geometric Applications of the Integral

1. **Question:** Find the volume of the solid formed by rotating the curve $y = 4$ around the x axis from $x = 6$ to $x = 10$.

   Solution: 64π or 201.0618

   Explanation: We will use Equation 20.2 to solve this:

   $$\text{volume} = \int_{x=\text{start}}^{x=\text{finish}} \pi y^2 \, dx$$

   $$= \int_6^{10} \pi 4^2 dx$$

   $$= \int_6^{10} 16\pi dx$$

   $$= \int 16\pi dx \Big|_6^{10}$$

   $$= 16\pi x \Big|_6^{10}$$

   $$= 16\pi(10) - 16\pi(6)$$

   $$= 64\pi$$

   $$\approx 201.0618$$

2. **Question:** Find the volume of the solid formed by rotating the curve $y = x$ around the y axis from $x = 1$ to $x = 6$.

   Solution: $\frac{430\pi}{3}$ or 450.2946 cubic units

   Explanation: We will use Equation 20.3

to solve this:

   $$\text{volume} = \int_{x=\text{start}}^{x=\text{finish}} 2\pi x y \, dx$$

   $$= \int_1^6 2\pi x x \, dx$$

   $$= \int_1^6 2\pi x^2 \, dx$$

   $$= \int 2\pi x^2 \, dx \Big|_1^6$$

   $$= \frac{2\pi}{3} x^3 \Big|_1^6$$

   $$= \frac{2\pi}{3}(6)^3 - \frac{2\pi}{3}(1)^3$$

   $$= \frac{432\pi}{3} - \frac{2\pi}{3}$$

   $$= \frac{430\pi}{3}$$

   $$\approx 450.2946$$

3. **Question:** Find the volume of the solid formed by rotating the curve $5y = x$ around the x axis from $x = 2$ to $x = 5$.

   Solution: $\frac{39\pi}{25}$ or 4.9009 cubic units

   Explanation: We will use Equation 20.2 to solve this. However, that equation requires a replacement for y itself. Therefore, we will first have to solve for y:

   $$5y = x$$

   $$y = \frac{x}{5}$$

Now we can apply Equation 20.2:

$$\text{volume} = \int_{x=\text{start}}^{x=\text{finish}} \pi y^2 \, dx$$

$$= \int_2^5 \pi \left(\frac{x}{5}\right)^2 dx$$

$$= \int_2^5 \pi \frac{x^2}{25} \, dx$$

$$= \int \pi \frac{x^2}{25} \, dx \Big|_2^5$$

$$= \pi \frac{x^3}{75} \Big|_2^5$$

$$= \pi \frac{(5)^3}{75} - \pi \frac{(2)^3}{75}$$

$$= \pi \frac{125}{75} - \pi \frac{8}{75}$$

$$= \frac{117\pi}{75}$$

$$= \frac{39\pi}{25}$$

$$\approx 4.9009$$

4. **Question:** Find the volume of the solid formed when the curve $y = \frac{1}{x}$ is rotated around the x axis from $x = 1$ to $x = 3$.

Solution: The volume is $\frac{2}{3}\pi$ or 2.0944 cubic units.

Explanation: We will use Equation 20.2

to solve this:

$$\text{volume} = \int_{x=\text{start}}^{x=\text{finish}} \pi y^2 \, dx$$

$$= \int_1^3 \pi \left(\frac{1}{x}\right)^2 dx$$

$$= \int \pi \left(\frac{1}{x}\right)^2 dx \Big|_1^3$$

$$= \pi \int x^{-2} \, dx \Big|_1^3$$

$$= \pi \left(-\frac{1}{x}\right) \Big|_1^3$$

$$= \pi \left(-\frac{1}{3} - -\frac{1}{1}\right)$$

$$= \pi \left(-\frac{1}{3} + 1\right)$$

$$= \frac{2}{3}\pi$$

$$\approx 2.0944$$

5. **Question:**   Find the length of the curve $y = \frac{x^3}{12} + \frac{1}{x}$ from $x = 1$ to $x = 2$.

Solution: The length of the curve is $\frac{13}{12}$ units long.

Explanation: We will use Equation 20.4 to solve this. In order to use that equation, we need to use differentials. Therefore, we will need to differentiate our original equation:

$$y = \frac{x^3}{12} + \frac{1}{x}$$

$$dy = \frac{3x^2}{12} \, dx + -\frac{1}{x^2} \, dx$$

$$dy = \frac{x^2}{4} \, dx + -\frac{1}{x^2} \, dx$$

Now, we will use this equivalency to get Equation 20.4 in terms of a single differential and then simplify it a bit:

$$\text{length} = \int_{x=\text{start}}^{x=\text{finish}} \sqrt{dx^2 + dy^2}$$

$$= \int_1^2 \sqrt{dx^2 + \left(\frac{x^2}{4}\,dx + -\frac{1}{x^2}\,dx\right)^2}$$

$$= \int_1^2 \sqrt{dx^2 + \frac{x^4}{16}\,dx^2 + 2\left(-\frac{1}{x^2}\right)\left(\frac{x^2}{4}\right)dx^2 + \frac{1}{x^4}\,dx^2}$$

$$= \int_1^2 \sqrt{dx^2\left(1 + \frac{x^4}{16} + 2\left(-\frac{1}{x^2}\right)\left(\frac{x^2}{4}\right) + \frac{1}{x^4}\right)}$$

$$= \int_1^2 \sqrt{\left(1 + \frac{x^4}{16} - \frac{1}{2} + \frac{1}{x^4}\right)}\,dx$$

This is still an ugly integral, but it is much cleaner if we find a common denominator to make it into a single fraction:

$$\text{length} = \int_1^2 \sqrt{\left(\frac{16x^4}{16x^4} + \frac{x^8}{16x^4} - \frac{8x^4}{16x^4} + \frac{16}{16x^4}\right)}\,dx$$

$$= \int_1^2 \sqrt{\frac{16x^4 + x^8 - 8x^4 + 16}{16x^4}}\,dx$$

$$= \int_1^2 \sqrt{\frac{x^8 + 8x^4 + 16}{16x^4}}\,dx$$

Interestingly, the top and the bottom of this fraction are both squares! This will enable us to get rid of our square root:

$$\text{length} = \int_1^2 \sqrt{\frac{(x^4 + 4)^2}{(4x^2)^2}}\,dx$$

$$= \int_1^2 \frac{\sqrt{(x^4 + 4)^2}}{\sqrt{(4x^2)^2}}\,dx$$

$$= \int_1^2 \frac{x^4 + 4}{4x^2}\,dx$$

$$= \int_1^2 \frac{x^4}{4x^2}\,dx + \frac{4}{4x^2}\,dx$$

$$= \int_1^2 \frac{x^2}{4}\,dx + x^{-2}\,dx$$

Now we are ready to solve the integral using basic polynomial rules:

$$\text{length} = \int_1^2 \frac{x^2}{4}\,dx + x^{-2}\,dx$$

$$= \int \frac{x^2}{4}\,dx + x^{-2}\,dx \Big|_1^2$$

$$= \frac{x^3}{12} + -\frac{1}{x} \Big|_1^2$$

$$= \frac{(2)^3}{12} + -\frac{1}{2} - \left(\frac{(1)^3}{12} + -\frac{1}{1}\right)$$

$$= \frac{8}{12} - \frac{6}{12} - \frac{1}{12} + \frac{12}{12}$$

$$= \frac{13}{12}$$

The length of the curve is $\frac{13}{12}$ units long.

6. **Question:** Find the volume of the solid formed when the curve $y = e^x$ is rotated around the x axis from $x = 0$ to $x = 2$.

   Solution: The volume is approximately 84.1918 cubic units.

   Explanation: We will use Equation 20.2 to solve this:

   $$\text{volume} = \int_{x=\text{start}}^{x=\text{finish}} \pi y^2\,dx$$

   $$= \int_0^2 \pi(e^x)^2\,dx$$

   $$= \int_0^2 \pi e^{2x}\,dx$$

   $$= \int \pi e^{2x}\,dx \Big|_0^2$$

   $$= \frac{\pi}{2}e^{2x} \Big|_0^2$$

   $$= \frac{\pi}{2}e^{2(2)} - \frac{\pi}{2}e^{2(0)}$$

   $$= \frac{\pi}{2}e^4 - \frac{\pi}{2}$$

   $$\approx 85.7626 - 1.5708$$

   $$\approx 84.1918$$

7. **Question:** Find the volume of the solid formed when the curve $y = x^3$ is rotated around the y

axis from $x = 0$ to $x = 3$.

Solution: The volume is $\frac{486\pi}{5}$ or 305.3625 cubic units.

Explanation: We will use Equation 20.3 to solve this:

$$\text{volume} = \int_{x=\text{start}}^{x=\text{finish}} 2\pi x y \, dx$$

$$= \int_{0}^{3} 2\pi x x^3 \, dx$$

$$= \int_{0}^{3} 2\pi x^4 \, dx$$

$$= \int 2\pi x^4 \, dx \Big|_{0}^{3}$$

$$= \frac{2\pi}{5} x^5 \Big|_{0}^{3}$$

$$= \frac{2\pi}{5}(3)^5 - \frac{2\pi}{5}(0)^5$$

$$= \frac{486\pi}{5} - 0$$

$$= \frac{486\pi}{5}$$

$$\approx 305.3625$$

8. **Question:** Find the length of the curve $y = \sqrt{4 - x^2}$ from $x = 0$ to $x = 2$. This will require you to consult Appendix H.7 for integral tables. It will also require some creative manipulation to complete.

Solution: π or 3.1416

Explanation: We will use Equation 20.4 to solve this. However, in order to do so, we will need to differentiate the original equation to find an equation in terms of differentials:

$$y = \sqrt{4 - x^2}$$

$$dy = \frac{1}{2}(4 - x^2)^{-\frac{1}{2}} \cdot (-2x) \, dx$$

$$dy = -x(4 - x^2)^{-\frac{1}{2}} \, dx$$

Now we can use Equation 20.4:

$$\text{length} = \int_{x=\text{start}}^{x=\text{finish}} \sqrt{dx^2 + dy^2}$$

$$= \int_{0}^{2} \sqrt{dx^2 + (-x(4 - x^2)^{-\frac{1}{2}} \, dx)^2}$$

$$= \int_{0}^{2} \sqrt{dx^2 + \frac{x^2}{4 - x^2} \, dx^2}$$

$$= \int_{0}^{2} \sqrt{dx^2(1 + \frac{x^2}{4 - x^2})}$$

$$= \int_{0}^{2} \sqrt{1 + \frac{x^2}{4 - x^2}} \, dx$$

Now, in order to make the equation easier to manipulate, we will choose a common denominator to allow us to add the 1 to the fraction. Since the denominator of the fraction is $4 - x^2$, we can replace 1 with $\frac{4-x^2}{4-x^2}$. Then the equation will become:

$$\text{length} = \int_{0}^{2} \sqrt{\frac{4 - x^2}{4 - x^2} + \frac{x^2}{4 - x^2}} \, dx$$

$$= \int_{0}^{2} \sqrt{\frac{4 - x^2 + x^2}{4 - x^2}}$$

$$= \int_{0}^{2} \sqrt{\frac{4}{4 - x^2}} \, dx$$

That was a lot of work! But now it is almost ready to solve. If you look at the form, it almost looks like the form that integrates to arcsin, but we need to massage it a little more first.

$$\text{length} = \int_{0}^{2} \sqrt{\frac{4}{4 - x^2}} \, dx$$

$$= \int \sqrt{\frac{4}{4 - x^2}} \, dx \Big|_{0}^{2}$$

$$= \int \sqrt{\frac{4}{4\left(1 - \left(\frac{x}{2}\right)^2\right)}} \, dx \Big|_{0}^{2}$$

$$= \int \sqrt{\frac{1}{1 - \left(\frac{x}{2}\right)^2}} \, dx \Big|_{0}^{2}$$

Now we need to use u substitution to finish it out. We will set $u = \frac{x}{2}$, so $du = \frac{dx}{2}$, or $2 \, du = dx$.

This gives us:

$$\text{length} = \int \sqrt{\frac{1}{1-u^2}}\, 2\, du \,\Big|_{x=0}^{x=2}$$

$$= \int 2\frac{1}{\sqrt{1-u^2}}\, du \,\Big|_{x=0}^{x=2}$$

$$= 2\arcsin(u) \,\Big|_{x=0}^{x=2}$$

$$= 2\arcsin\left(\frac{x}{2}\right) \,\Big|_{x=0}^{x=2}$$

$$= 2(\arcsin(1) - \arcsin(0))$$

$$= 2\frac{\pi}{2}$$

$$= \pi$$

$$\approx 3.1416$$

Also, if you spend some time thinking about it, you might be able to come up with another (non-calculus) way to determine that the length of this line is π.

Chapter 21

Transforming Equations for Integration

Integrate the following using partial fraction decomposition:

1. **Question:** $\int \frac{2x+3}{x^2-x-2}\,dx$

   Solution: $\frac{7}{3}\ln|x-2| + \frac{-1}{3}\ln|x+1| + C$

   Explanation: To solve this, first we have to factor the denominator. The denominator factors into $(x-2)(x+1)$. Therefore, we say that:
   $$\frac{2x+3}{x^2-x-2} = \frac{A}{x-2} + \frac{B}{x+1}$$
   Putting the fractions over a common denominator yields:
   $$\frac{A(x+1) + B(x-2)}{x^2-x-2} = \frac{Ax + A + Bx - 2B}{x^2-x-2}$$
   The numerator of the original fraction was $2x+3$. Therefore we can say that:
   $$Ax + A + Bx - 2B = 2x + 3$$
   This gives the following equations:
   $$Ax + Bx = 2x$$
   $$A - 2B = 3$$
   The first one can be easily solved for A as $A = 2 - B$. Therefore, the second one becomes:
   $$(2 - B) - 2B = 3$$
   $$2 - 3B = 3$$
   $$-3B = 1$$
   $$B = \frac{-1}{3}$$
   Therefore, $A = \frac{7}{3}$. We can rewrite the integral as:
   $$\int \frac{\frac{7}{3}}{x-2}\,dx + \frac{\frac{-1}{3}}{x+1}\,dx$$

This integrates to:
$$\frac{7}{3}\ln|x-2| + \frac{-1}{3}\ln|x+1| + C$$

2. **Question:** $\int \frac{x-5}{x^2-x}\,dx$

   Solution: $5\ln|x| - 4\ln|x-1| + C$

   Explanation: To solve this, firs twe have to factor the denominator. The denominator factors into $x(x-1)$. Therefore, we say that
   $$\frac{x-5}{x^2-x} = \frac{A}{x} + \frac{B}{x-1}$$
   Putting the fractions over a common denominator yields:
   $$\frac{A(x-1) + B(x)}{x^2-x} = \frac{Ax - A + Bx}{x^2-x}$$
   The numerator of the original fraction was $x-5$. Therefore we can say that:
   $$Ax - A + Bx = x - 5$$
   This gives the following equations:
   $$Ax + Bx = x$$
   $$-A = -5$$
   The second one is easier—we can just say that $A = 5$. Substituting into the first equation gives us:
   $$5 + B = 1$$
   $$B = -4$$
   Therefore, we can rewrite the integral as:
   $$\int \frac{5}{x}\,dx + \frac{-4}{x-1}\,dx$$

This integrates to:

$$5\ln|x| - 4\ln|x-1| + C$$

3. **Question:** $\int \frac{x+4}{x^2+10x+25}\, dx$

Solution: $\ln|x+5| + \frac{1}{x+5} + C$

Explanation: To solve this, first we have to factor the denominator. The denominator factors into $(x+5)(x+5)$. This is a little different than usual, because the factor is squared. Therefore, this is handled by separating out into the following:

$$\frac{A}{x+5} + \frac{B}{(x+5)^2}$$

Putting the fractions over a common denominator yields:

$$\frac{Ax + 5A + B}{x^2 + 10x + 25}$$

The numerator of the original fraction was $x+4$. Therefore we can say that:

$$Ax + 5A + B = x + 4$$

This gives the following equations:

$$Ax = x$$

$$5A + B = 4$$

The first equation simplifies to just $A = 1$. Therefore, the second equation becomes:

$$5 + B = 4B = -1$$

Therefore, we can rewrite the integral as:

$$\int \frac{1}{x+5}\, dx + \frac{-1}{(x+5)^2}\, dx$$

This can be rewritten as:

$$\int \frac{1}{x+5}\, dx - (x+5)^{-2}\, dx$$

This integrates to:

$$\ln|x+5| + (x+5)^{-1} + C$$

Or, more naturally:

$$\ln|x+5| + \frac{1}{x+5} + C$$

4. **Question:** $\int \frac{x-12}{x^2-2x-35}\, dx$

Solution: $\frac{-5}{12}\ln|x-7| + \frac{17}{12}\ln|x+5| + C$

Explanation: To solve this, first we have to factor the denominator. The denominator factors into $(x-7)(x+5)$. Therefore, we say that:

$$\frac{x-12}{x^2-2x-35} = \frac{A}{x-7} + \frac{B}{x+5}$$

Putting the fractions over a common denominator yields:

$$\frac{A(x+5) + B(x-7)}{x^2-2x-35} = \frac{Ax + 5A + Bx - 7B}{x^2-2x-35}$$

The numerator of the original fraction as $x-12$. Therefore, we can say that:

$$Ax + 5A + Bx - 7B = x - 12$$

This gives the following equations:

$$Ax + Bx = x$$

$$5A - 7B = -12$$

The first one can be solved for A as $A = 1 - B$. Therefore, the second one becomes:

$$5(1-B) - 7B = -12$$

$$5 - 5B - 7B = -12$$

$$5 - 12B = -12$$

$$-12B = -17$$

$$B = \frac{17}{12}$$

Therefore, $A = \frac{-5}{12}$. We can rewrite the integral as:

$$\int \frac{\frac{-5}{12}}{x-7}\, dx + \frac{\frac{17}{12}}{x+5}\, dx$$

This integrates to:

$$\frac{-5}{12}\ln|x-7| + \frac{17}{12}\ln|x+5| + C$$

Integrate the following using trigonometric substitutions. Note that these problems may take a lot more creativity than previous homework assignments.

5. **Question:** $\int \frac{1}{\sqrt{1-x^2}}\, dx$

Solution: $\arcsin(x) + C$

Explanation: This matches one of the forms given for a trigonometric substitution. The suggestion from the chapter is that we us $x = \sin(u)$. Therefore, $dx = \cos(u)\, du$. Substituting this in for x yields:

$$\int \frac{1}{\sqrt{1-\sin^2(u)}} \cos(u)\, du$$

We can make use of the identity $\cos^2(u) = 1 - \sin^2(u)$ to simplify the expression:

$$\int \frac{1}{\sqrt{\cos^2(u)}} \cos(u)\, du$$

Now we can pull it out from under the square root sign:

$$\int \frac{1}{\cos(u)} \cos(u)\, du$$

This can be simplified quite a bit by cancelling out $\cos(u)$:

$$\int du$$

This integrates to simply:

$$u + C$$

Now, since $x = \sin(u)$, then that means that $u = \arcsin(x)$. So this becomes:

$$\arcsin(x) + C$$

6. **Question:** $\int \frac{\cos^2(x)}{1+\sin(x)}\, dx$

Solution: $x + \cos(x) + C$

Explanation: If you look at the numerator, you might remember that $\cos^2(x) = 1 - \sin^2(x)$. Why is that important? Because $1 - \sin^2(x)$ can be factored, with one of the factors matching

the denominator. Let's take a look:

$$\int \frac{\cos^2(x)}{1+\sin(x)}\, dx$$

$$= \int \frac{1 - \sin^2(x)}{1+\sin(x)}\, dx$$

$$= \int \frac{(1-\sin(x))(1+\sin(x))}{1+\sin(x)}\, dx$$

$$= \int (1 - \sin(x))\, dx$$

$$= \int dx - \sin(x)\, dx$$

$$= x + \cos(x) + C$$

7. **Question:** $\int \frac{1}{4+x^2}\, dx$

Solution: $\frac{1}{4} \arctan\left(\frac{x}{2}\right) + C$

Explanation: To solve this, we will make a substitution of $x = 2\tan(u)$. Therefore, $dx = \sec^2(u)\, du$ (from Appendix H.6). This makes our expression become:

$$\int \frac{1}{4} \frac{\sec^2(u)}{1+\tan^2(u)}\, du$$

Appendix H.5 tells us that $1 + \tan^2(u) = \sec^2(u)$. Therefore, the integral becomes:

$$\int \frac{1}{4} \frac{\sec^2(u)}{\sec^2(u)}\, du$$

This reduces to:

$$\int \frac{1}{4} du$$

This integrates to:

$$\frac{1}{4} u + C$$

Since $x = 2\tan(u)$, therefore, $u = \arctan\left(\frac{x}{2}\right)$. Therefore, the final answer is:

$$\frac{1}{4} \arctan\left(\frac{x}{2}\right) + C$$

8. **Question:** $\int \frac{1}{x^2 \sqrt{16-x^2}}\, dx$

   Solution: $-\frac{1}{4}\frac{\sqrt{1-\left(\frac{x}{4}\right)^2}}{x} + C$

   Explanation: To solve this problem, we are going to set $x = 4\sin(u)$. Therefore, $dx = 4\cos(u)\, du$. This makes the integral become:

$$\int \frac{1}{16\sin^2(u)\sqrt{16-16\sin^2(u)}}\, 4\cos(u)\, du$$

This can be simplified to:

$$\int \frac{4}{16}\frac{\cos(u)}{\sin^2(u)\sqrt{4^2(1-\sin^2(u))}}\, du$$

Since $1-\sin^2(u) = \cos^2(u)$ we can rewrite this as:

$$\int \frac{4}{16}\frac{\cos(u)}{\sin^2(u)\sqrt{4^2\cos^2(u)}}\, du$$

We can move our things out of the square root sign:

$$\int \frac{4}{16}\frac{\cos(u)}{4\cos(u)\sin^2(u)}\, du$$

Now we can cancel a lot of stuff:

$$\int \frac{1}{16}\frac{1}{\sin^2(u)}\, du$$

Since $\frac{1}{\sin(u)} = \csc(u)$, this becomes:

$$\int \frac{1}{16}\csc^2(u)\, du$$

The integral of $\csc^2(u)$ is given in Appendix H.7 as being $-\cot(u)$. Therefore, this integrates to:

$$-\frac{1}{16}\cot(u) + C$$

Since $x = 4\sin(u)$, $u = \arcsin\left(\frac{x}{4}\right)$. Therefore, substituting back in gives us:

$$-\frac{1}{16}\cot\left(\arcsin\left(\frac{x}{4}\right)\right) + C$$

Note that it is entirely legitimate to stop here, as we have the answer in terms of x.

However, because we have both trig functions and inverse trig functions, it can be helpful to unwrap it a little further.

cot is cotangent, which is just $\frac{\cos()}{\sin()}$. Therefore, this can be transformed into:

$$-\frac{1}{16}\frac{\cos\left(\arcsin\left(\frac{x}{4}\right)\right)}{\sin\left(\arcsin\left(\frac{x}{4}\right)\right)} + C$$

We can immediately simplify the denominator like this:

$$-\frac{1}{16}\frac{\cos\left(\arcsin\left(\frac{x}{4}\right)\right)}{\frac{x}{4}} + C$$

Taking the cosine of the arcsine takes a little more work. Remember that $\cos^2(u) = 1-\sin^2(u)$. Therefore, square rooting both sides gives us $\cos(u) = \sqrt{1-\sin^2(u)}$. Using that principle, we can transform this into:

$$-\frac{1}{16}\frac{\sqrt{1-\sin^2\left(\arcsin\left(\frac{x}{4}\right)\right)}}{\frac{x}{4}} + C$$

This reduces to:

$$-\frac{1}{16}\frac{\sqrt{1-\left(\frac{x}{4}\right)^2}}{\frac{x}{4}} + C$$

We can simplify it a little further into this:

$$-\frac{1}{4}\frac{\sqrt{1-\left(\frac{x}{4}\right)^2}}{x} + C$$

This is the final answer.

Chapter 22

Integration by Parts

Integral the following:

1. **Question:** $\int x \sin(x)\mathrm{d}x$

   Solution: $\sin(x) - x\cos(x) + C$

   Explanation: Since x is the term progressing towards $\mathrm{d}x$ and $\sin(x)$ is a cyclical derivative, we will use x for u and $\sin(x)\,\mathrm{d}x$ for $\mathrm{d}v$. You can easily see that $\mathrm{d}u = \mathrm{d}x$ and $v = -\cos(x)$. Therefore, starting with the integration by parts formula, we can do:

   $$\int x \sin(x)\mathrm{d}x = -x\cos(x) - \int -\cos(x)\,\mathrm{d}x$$

   $$= -x\cos(x) + \int \cos(x)\,\mathrm{d}x$$

   $$= -x\cos(x) + \sin(x) + C$$

   $$= \sin(x) - x\cos(x) + C$$

2. **Question:** $\int x\, e^x \mathrm{d}x$

   Solution: $x\, e^x - e^x + C$

   Explanation: Since x is the term progressing towards $\mathrm{d}x$ and e^x is a cyclical derivative, we will use x for u and $e^x \mathrm{d}x$ for $\mathrm{d}v$. You can see that $\mathrm{d}u = \mathrm{d}x$ and $v = e^x$. Therefore, starting with the integration by parts formula, we can do:

   $$\int x e^x \mathrm{d}x = x e^x - \int e^x \mathrm{d}x$$

   $$= x e^x - e^x + C$$

3. **Question:** $\int x^2 3^x \mathrm{d}x$

   Solution:

   $$\frac{1}{\ln(3)}x^2 3^x - \frac{2}{\ln(3)}\left(\frac{1}{\ln(3)}x\,3^x - \frac{1}{\ln(3)}\frac{1}{\ln(3)}3^x\right) + C$$

   There are many equivalent simplifications of this available as well. One possible simplification is:

   $$\frac{x^2\,3^x}{\ln(3)} - \frac{2x\,3^x}{(\ln(3))^2} + \frac{2\cdot 3^x}{(\ln(3))^3} + C$$

   Explanation: In this problem, we have a term (x^2) whose derivatives are progressing towards $\mathrm{d}x$, but it will take two steps (i.e., we will have to do integration by parts twice). 3^x does integrate in a circle, with the exception that it keeps generating constant divisors. However, since they are constant, they won't stand in our way.

   So, we will start by saying $u = x^2$ and $\mathrm{d}v = 3^x \mathrm{d}x$. That means that $\mathrm{d}u = 2x\,\mathrm{d}x$ and $v = \frac{1}{\ln(3)}3^x$. Using integration by parts, we can see:

   $$\int x^2 3^x \mathrm{d}x = \frac{1}{\ln(3)}x^2 3^x - \int \frac{1}{\ln(3)}3^x \cdot 2x\,\mathrm{d}x$$

   $$= \frac{1}{\ln(3)}x^2 3^x - \frac{2}{\ln(3)}\int x\,3^x \mathrm{d}x$$

   The integral on the rightmost side, $\int x\,3^x \mathrm{d}x$ can be solved using integration by parts again. We will set $u = x$ and $\mathrm{d}v = 3^x \mathrm{d}x$. Then, $\mathrm{d}u = \mathrm{d}x$ and

$v = \frac{1}{\ln(3)} 3^x$. Therefore:

$$\int x\, 3^x\, dx = \frac{1}{\ln(3)} x\, 3^x - \int \frac{1}{\ln(3)} 3^x\, dx$$

$$= \frac{1}{\ln(3)} x\, 3^x - \frac{1}{\ln(3)} \int 3^x\, dx$$

$$= \frac{1}{\ln(3)} x\, 3^x - \frac{1}{\ln(3)} \frac{1}{\ln(3)} 3^x$$

Now, we just need to put all of the parts back together. This will give us a final expression of:

$$\frac{1}{\ln(3)} x^2\, 3^x - \frac{2}{\ln(3)} \left(\frac{1}{\ln(3)} x\, 3^x - \frac{1}{\ln(3)} \frac{1}{\ln(3)} 3^x \right)$$

If you like, this can be simplified to:

$$\frac{x^2\, 3^x}{\ln(3)} - \frac{2x\, 3^x}{(\ln(3))^2} + \frac{2 \cdot 3^x}{(\ln(3))^3}$$

4. **Question:** $\int 2x^2 \cos(x)\, dx$

Solution: $2x^2 \sin(x) + 4x \cos(x) - 4 \sin(x) + C$

Explanation: In this integral, we have a product of functions. This should make us think of solving it using integration by parts. The first step is to assign which parts of the term are in u and which parts are in dv.

Notice that differentials of $2x^2$ would eventually become just dx, and that $\cos(x)$ integrates in a circle. Therefore, we will set $u = 2x^2$ and $dv = \cos(x)\, dx$. This means that $du = 4x\, dx$ and $v = \sin(x)$ Therefore, the integral can be rewritten as:

$$\int 2x^2 \cos(x)\, dx = 2x^2 \sin(x) - \int 4x \sin(x)\, dx$$

The integral on the right is easier, but we will need to do integration by parts again.

In the integral $\int 4x \sin(x)\, dx$, the $4x$ is progressing towards dx. Therefore, we will set $u = 4x$ and $dv = \sin(x)\, dx$. This means that $du = 4\, dx$ and $v = -\cos(x)$. Integration by parts gives us:

$$\int 4x \sin(x)\, dx = -4x \cos(x) - \int -4 \cos(x)\, dx$$

$\int -4 \cos(x)\, dx$ can be easily integrated into $-4 \sin(x)$. That means that:

$$\int 4x \sin(x)\, dx = -4x \cos(x) + 4 \sin(x)$$

Plugging this back into our earlier equation, we can see that:

$$\int 2x^2 \cos(x)\, dx$$

$$= 2x^2 \sin(x) - (-4x \cos(x) + 4 \sin(x))$$

$$= 2x^2 \sin(x) + 4x \cos(x) - 4 \sin(x)$$

$$= 2x^2 \sin(x) + 4x \cos(x) - 4 \sin(x) + C$$

5. **Question:** $\int x \ln(x)\, dx$

Solution: $\frac{1}{2} x^2 \ln(x) - \frac{1}{4} x^2 + C$

Explanation: If you think about these two terms, differentiating x will lead to just dx while differentiating $\ln(x)$ will lead to $\frac{dx}{x}$. Therefore, we will choose x for our u and $\ln(x)\, dx$ for our dv. We will need to look up the integral of $\ln(x)\, dx$ from Appendix H.7. This gives us, $du = dx$ and $v = \int \ln(x)\, dx = x \ln(x) - x$.

Using integration by parts, this means that:

$$\int x \ln(x)\, dx = x(x \ln(x) - x) - \left(\int x \ln(x)\, dx - x\, dx \right)$$

Note that, on the right-hand side, we have another instance of $x \ln(x)\, dx$. Therefore, we just need to separate it out and add it to both sides. To do this, we will separate the integral into two pieces at the point of subtraction, which will yield:

$$\int x \ln(x)\, dx = x(x \ln(x) - x) - \int x \ln(x)\, dx + \int x\, dx$$

If we add $\int x \ln(x)\, dx$ to both sides, that will yield:

$$2 \int x \ln(x)\, dx = x(x \ln(x) - x) + \int x\, dx$$

That integral is easy—just the power rule:

$$2 \int x \ln(x)\, dx = x(x \ln(x) - x) + \frac{1}{2} x^2$$

Now we can divide both sides by two:

$$\int x \ln(x) \, dx = \frac{1}{2}x(x \ln(x) - x) + \frac{1}{4}x^2$$

We can also distribute terms for a more simplified answer:

$$\int x \ln(x) \, dx = \frac{1}{2}x^2 \ln(x) - \frac{1}{2}x^2 + \frac{1}{4}x^2$$

Taking the x^2 terms together gives us the following result:

$$\int x \ln(x) \, dx = \frac{1}{2}x^2 \ln(x) - \frac{1}{4}x^2$$

Adding in the constant of integration will give us the final answer:

$$\int x \ln(x) \, dx = \frac{1}{2}x^2 \ln(x) - \frac{1}{4}x^2 + C$$

6. **Question:** $\int 3x \, 5^{2x} \, dx$

Solution: $\frac{3x \, 5^{2x}}{2 \ln(5)} - \frac{3 \cdot 5^{2x}}{4 \ln^2(5)} + C$

Explanation: In this case, the differential of $3x$ will leave us with a simple differential, and 5^{2x}, while not integrating *exactly* in a circle, is close enough (the offset is by a constant factor, not a variable factor). Therefore, we will set $u = 3x$ and $dv = 5^{2x} \, dx$. That means that $du = 3 \, dx$ and $v = \frac{5^{2x}}{2 \ln(5)}$. Using integration by parts allows us to write out the following equation:

$$\int 3x \, 5^{2x} \, dx = 3x \frac{5^{2x}}{2 \ln(5)} - \int \frac{5^{2x}}{2 \ln(5)} 3 \, dx$$

We can move all of the constants out in front of the integral to simplify it. Then it becomes:

$$\int 3x \, 5^{2x} \, dx = 3x \frac{5^{2x}}{2 \ln(5)} - \frac{3}{2 \ln(5)} \int 5^{2x} \, dx$$

$\int 5^{2x} \, dx$ integrates very simply to $\frac{5^{2x}}{2 \ln(5)}$. Therefore, this becomes:

$$\int 3x \, 5^{2x} \, dx = 3x \frac{5^{2x}}{2 \ln(5)} - \frac{3}{2 \ln(5)} \frac{5^{2x}}{2 \ln(5)}$$

This can be simplified to:

$$\int 3x \, 5^{2x} \, dx = \frac{3x \, 5^{2x}}{2 \ln(5)} - \frac{3 \cdot 5^{2x}}{4 \ln^2(5)}$$

Adding in the constant of integration yields:

$$\int 3x \, 5^{2x} \, dx = \frac{3x \, 5^{2x}}{2 \ln(5)} - \frac{3 \cdot 5^{2x}}{4 \ln^2(5)} + C$$

There are actually quite a number of ways to write this, but this is sufficient for our purposes.

Integrate the following without using the table of integrals in Appendix H.7. You *may* use the table of differentials in Appendix H.6. Also note that these may require more than one technique to perform.

7. **Question:** $\int \ln(x) \, dx$

Solution: $x \ln(x) - x + C$

Explanation: To solve this integral, we will set $u = \ln(x)$ and $dv = dx$. Therefore, $du = \frac{dx}{x}$ and $v = x$. Using integration by parts means that we can write the integral as:

$$\int \ln(x) \, dx = x \ln(x) - \int x \frac{1}{x} \, dx$$

The right-hand side simplifies quite a bit because x and $\frac{1}{x}$ cancel out. This gives us:

$$\int \ln(x) \, dx = x \ln(x) - \int dx$$

This can be easily solved to:

$$\int \ln(x) \, dx = x \ln(x) - x + C$$

8. **Question:** $\int \arcsin(x) \, dx$

Solution: $x \arcsin(x) + \sqrt{1 - x^2} + C$

Explanation: Since we have a differential for $\arcsin(u)$, we can use that to try to do an integration by parts to integrate arcsin. So, we will start by setting up the integration by parts formula, and set $u = \arcsin(x)$ and $dv = dx$. Therefore, $v = x$ and $du = \frac{dx}{\sqrt{1-x^2}}$. Therefore, the integration by parts formula tells us:

$$\int \arcsin(x) \, dx = x \arcsin(x) - \int \frac{x}{\sqrt{1 - x^2}} \, dx$$

Now, we are going to put aside part of this equation, and focus on the remaining integral. We will put it all back together at the end. For now, though, we will work with this:

$$\int \frac{x}{\sqrt{1-x^2}}\, dx$$

Now, from Chapter 21 we know that, since we have $\sqrt{1-x^2}$ we can do a u substitution of $x = \sin(u)$ for this. If $x = \sin(u)$ then $dx = \cos(u)\, du$. This makes the integral become:

$$\int \frac{\sin(u)\,\cos(u)\, du}{\sqrt{1-\sin^2(u)}}$$

Using the relationship $\cos^2(u) = 1 - \sin^2(u)$, this becomes:

$$\int \frac{\sin(u)\,\cos(u)\, du}{\sqrt{\cos^2(u)}}$$

The square root of $\cos^2(u)$ is just $\cos(u)$, so this becomes:

$$\int \frac{\sin(u)\,\cos(u)\, du}{\cos(u)}$$

The $\cos(u)$ on the top and the bottom cancel out, which gives us just:

$$\int \sin(u)\, du$$

This integrates to:

$$-\cos(u)$$

However, we need to back-substitute for u. Therefore, since $x = \sin(u)$, then $u = \arcsin(x)$. That makes the result become:

$$-\cos(\arcsin(x))$$

Now, we can replace $\cos()$ with $\sqrt{1 - sin^2()}$. Therefore, this becomes:

$$-\sqrt{1 - \sin^2(\arcsin(x))}$$

The sin of $\arcsin(x)$ is just the x, so this becomes:

$$-\sqrt{1-x^2}$$

This, then, replaces the integral in the right-hand side of the following equation:

$$\int \arcsin(x) dx = x\,\arcsin(x) - \int \frac{x}{\sqrt{1-x^2}}\, dx$$

Therefore, this becomes:

$$\int \arcsin(x) dx = x\,\arcsin(x) - (-\sqrt{1-x^2})$$

Which simplifies to:

$$\int \arcsin(x) dx = x\,\arcsin(x) + \sqrt{1-x^2}$$

Now we add in the constant of integration, and we get:

$$\int \arcsin(x) dx = x\,\arcsin(x) + \sqrt{1-x^2} + C$$

This result could also be obtained by doing more basic u substitutions, and the same result is obtained. You can see that version of the proof in Appendix F.10 of the main book.

Solve the following:

9. **Question:** Find the volume of the solid formed when the curve $y = e^x$ is rotated around the y axis from $x = 0$ to $x = 2$.

Solution: $2\pi e^2 + 2\pi$ or 52.7100

Explanation: We will use Equation 20.3 to solve this:

$$\begin{aligned}
\text{volume} &= \int_{x=\text{start}}^{x=\text{finish}} 2\pi xy\, dx \\
&= \int_0^2 2\pi x e^x\, dx \\
&= \int 2\pi x\, e^x\, dx \Big|_0^2
\end{aligned}$$

To simplify this, we will move the constants outside of the integral, giving us:

$$2\pi \int x\, e^x\, dx \Big|_0^2$$

To simplify, we will remove the constants and the evaluation bar and just focus on the integral for the moment:

$$\int x\, e^x\, dx$$

To solve this, we need to use integration by parts. So we will set $u = x$ and $dv = e^x\, dx$. Therefore, $du = dx$ and $v = e^x$. Using the integration by parts formula, we can see that:

$$\int x\, e^x\, dx = x\, e^x - \int e^x\, dx$$

Additionally, $\int e^x \, dx$ integrates simply to e^x. Therefore, the integral becomes:

$$\int x e^x \, dx = x e^x - e^x$$

Since this is a definite integral, we don't need to worry about the constant of integration. However, we did shave off some constant multipliers at the front of the integral, as well as the evaluation bar. Therefore, the fuller result is:

$$\text{volume} = 2\pi (x e^x - e^x) \Big|_0^2$$

This can be evaluated in a fairly straightforward manner:

$$\text{volume} = 2\pi (x e^x - e^x) \Big|_0^2$$
$$= 2\pi((2)e^2 - e^2) - 2\pi(0e^0 - e^0)$$
$$= 2\pi(e^2) - 2\pi(-1)$$
$$= 2\pi e^2 + 2\pi$$
$$\approx 52.7100$$

Now, let's integrate both sides:

$$\int d\left(\frac{u}{v}\right) = \int \frac{v \, du}{v^2} - \int \frac{u \, dv}{v^2}$$

The left-hand side reduces since it is the integral of the differential:

$$\frac{u}{v} = \int \frac{v \, du}{v^2} - \int \frac{u \, dv}{v^2}$$

Then, we can solve for either integral. This will yield either:

$$\int \frac{v \, du}{v^2} = \frac{u}{v} + \int \frac{u \, dv}{v^2}$$

Or, if you solve for the other integral, you will get:

$$\int \frac{u \, dv}{v^2} = \int \frac{v \, du}{v^2} - \frac{u}{v}$$

10. **Question:** Create an "integration by parts" rule using the quotient rule (i.e., $d(\frac{u}{v})$). Note that you will need to split the fraction to create the rule.

Solution: Either of these are valid integration by parts rules:

$$\int \frac{v \, du}{v^2} = \frac{u}{v} + \int \frac{u \, dv}{v^2}$$

or

$$\int \frac{u \, dv}{v^2} = \int \frac{v \, du}{v^2} - \frac{u}{v}$$

Additionally, for both of them, $\int \frac{v \, du}{v^2}$ can be reduced to $\int \frac{du}{v}$ in your answer.

Explanation: The differential of $\frac{u}{v}$ is:

$$d\left(\frac{u}{v}\right) = \frac{v \, du - u \, dv}{v^2}$$

First, let's split the fraction so that it becomes:

$$d\left(\frac{u}{v}\right) = \frac{v \, du}{v^2} - \frac{u \, dv}{v^2}$$

Chapter 23

Problem-Solving Using the Integral

1. **Question:** The acceleration due to gravity is $-9.8067\frac{m}{sec^2}$ (meters per second squared). The acceleration is negative because acceleration brings you back to earth rather than up, up, and away. Derive an equation for position given the acceleration due to gravity. The initial velocity is given as v_0 and the initial position is given as p_0.

Solution: $-4.9034t^2 + v_0 t + p_0$

Explanation: Acceleration is the change in velocity over time, or, in other words, the differential $\frac{dv}{dt}$. Therefore, since the acceleration is given as $-9.8067\frac{m}{sec^2}$, we can write that as:

$$\frac{dv}{dt} = -9.8067\frac{m}{sec^2}$$

The actual velocity is the sum of accumulations of changes to the velocity, or, in other words, the integral:

$$\frac{dv}{dt} = -9.8067\frac{m}{sec^2}$$

$$dv = -9.8067\frac{m}{sec^2} dt$$

$$\int dv = \int -9.8067\frac{m}{sec^2} dt$$

$$v = -9.8067\frac{m}{sec^2}t + C$$

Remember, m is a meter, and sec is a second, which are constant units. Even though we don't have a *number* for them, they are still treated as constants since they don't change.

To find out what C should be, just substitute in 0 for the time, and the starting velocity for v:

$$v_0 = -9.8067\frac{m}{sec^2} \cdot 0 + C$$

$$v_0 = C$$

So, C just becomes v_0 in the equation:

$$v = -9.8067\frac{m}{sec^2}t + v_0$$

Now, velocity is merely the change in position over time, or $\frac{dp}{dt}$. In other words, the change in position is merely the sum (or accumulation) of all of the individual changes in position given by the velocity. So, we can find position by integrating velocity.

$$\frac{dp}{dt} = -9.8067\frac{m}{sec^2}t + v_0$$

$$dp = -9.8067\frac{m}{sec^2}t\, dt + v_0\, dt$$

$$\int dp = \int -9.8067\frac{m}{sec^2}t\, dt + v_0\, dt$$

$$p = -9.8067\frac{m}{sec^2}\frac{t^2}{2} + v_0 t + C$$

$$p \approx -4.9034\frac{m}{sec^2}t^2 + v_0 t + C$$

So, since p_0 is the starting position, we can solve for C by substitutinng in $t = 0$ and $p = p_0$ into the equation:

$$p = -4.9034\frac{m}{sec^2}t^2 + v_0 t + C$$

$$p_0 = -4.9034\frac{m}{sec^2}\frac{(0)^2}{+}v_0(0) + C$$

$$p_0 = 0 + 0 + C$$

$$p_0 = C$$

Therefore, the final equation becomes:

$$p = -4.9034\frac{m}{sec^2}t^2 + v_0 t + p_0$$

2. **Question:** Using the previous equation, if I dropped a ball from a tower 100 m tall, how long would it take to hit the ground? We will use the top of the tower as position 100 m, and the ground as position 0 m.

Solution: 4.5160 sec

Explanation: Since we are *dropping* the ball, the initial velocity (v_0) is 0. Since we are on the building, the initial position is 100 m. Therefore, we can reduce the equation as follows:

$$p = -4.9034 \frac{m}{sec^2} t^2 + v_0 t + p_0$$

$$p = -4.9034 \frac{m}{sec^2} t^2 + (0)t + 100 \, m$$

$$p = -4.9034 \frac{m}{sec^2} t^2 + 100 \, m$$

Now, when the ball hits the ground, p will be 0. So, we can substitute that in and solve for t:

$$p = -4.9034 \frac{m}{sec^2} t^2 + 100 \, m$$

$$0 = -4.9034 \frac{m}{sec^2} t^2 + 100 \, m$$

$$-100 \, m = -4.9034 \frac{m}{sec^2} t^2$$

$$-100 \, m \frac{1}{-4.9034} \frac{sec^2}{m} = t^2$$

$$\frac{-100}{-4.9034} \frac{m \, sec^2}{m} = t^2$$

$$\frac{-100}{-4.9034} sec^2 = t^2$$

$$20.3940 \, sec^2 \approx t^2$$

$$\sqrt{20.3940 \, sec^2} \approx \sqrt{t^2}$$

$$4.5160 \, sec \approx t$$

Therefore, the ball will hit the ground at approximately 4.5160 seconds after dropping it.

3. **Question:** Instead of a *constant* acceleration, let's say the equation for the acceleration of a vehicle at a given time is given by the formula $a = 3\sqrt{t}$. Assuming that position and velocity are zero when the time begins, at what position is the vehicle when $t = 5$.

Solution: The position of the vehicle will be 44.7214 units.

Explanation: Acceleration is the derivative of velocity, so $a = \frac{dv}{dt}$. Therefore, the formula $a = 3\sqrt{t}$ can be rewritten as:

$$\frac{dv}{dt} = 3\sqrt{t}$$

Additionally, we can rewrite $\sqrt{t}$ as $t^{0.5}$ which gives the equation:

$$\frac{dv}{dt} = 3t^{0.5}$$

We can integrate this to find an equation for v.

$$\frac{dv}{dt} = 3t^{0.5}$$

$$dv = 3t^{0.5} \, dt$$

$$\int dv = \int 3t^{0.5} \, dt$$

$$v = \frac{3t^{1.5}}{1.5} + C$$

$$v = 2t^{1.5} + C$$

Since the starting velocity is given as 0, when $t = 0$ then $v = 0$. Therefore, we can solve for C and see that $C = 0$ as well:

$$0 = 2(0)^{1.5} + C$$

Therefore, the equation for velocity is:

$$v = 2t^{1.5}$$

However, velocity is just the derivative of position, $v = \frac{dp}{dt}$. Therefore, we can use this idea to solve for position:

$$v = 2t^{1.5}$$

$$\frac{dp}{dt} = 2t^{1.5}$$

$$dp = 2t^{1.5} \, dt$$

$$\int dp = \int 2t^{1.5} \, dt$$

$$p = \frac{2t^{2.5}}{2.5} + C$$

$$p = 0.8t^{2.5} + C$$

Again, since the starting position is 0, then $C = 0$. So the equation for position becomes:

$$p = 0.8t^{2.5}$$

The question is for the position at $t = 5$. Therefore:

$$p = 0.8t^{2.5}$$
$$= 0.8(5)^{2.5}$$
$$\approx 0.8 \cdot 55.9017$$
$$\approx 44.7214$$

4. **Question:** If you deposit your money in a bank that has an interest rate of 4% per year, work out an equation, starting with differentials, of how much money (m) you will have after t years, if you start out with m_0 dollars. Use standard assumptions of calculus to find the equation.

Solution: $m = m_0\, e^{0.04t}$

Explanation: This is a compound accumulation problem. Remember that, for calculus, we presume that these sorts of things happen smoothly and continuously over time. So, you will be receiving money continually.

The amount of money you receive in interest at any instant is the change in money over time, $\frac{dm}{dt}$. Since the *rate* is 4%, then the rate in terms of actual dollars is $0.04m$. Therefore, we can say that:

$$\frac{dm}{dt} = 0.04m$$

We can solve this for an absolute amount of money at any given time by integrating:

$$\frac{dm}{dt} = 0.04m$$
$$\frac{dm}{m} = 0.04\,dt$$
$$\int \frac{dm}{m} = \int 0.04\,dt$$
$$\ln(m) = 0.04t + C$$
$$e^{\ln(m)} = e^{0.04t+C}$$
$$m = e^{0.04t+C}$$
$$m = e^C\, e^{0.04t}$$

Because e^C is just another arbitrary constant, we can reduce this to just C, giving us:

$$m = C\, e^{0.04t}$$

At time $t = 0$, $m = C$, which means that $m_0 = C$. Therefore, we can write this as:

$$m = m_0\, e^{0.04t}$$

5. **Question:** Find the average (mean) value of the function $\sin(x)$ from $x = 1$ to $x = 1.5$ (use radians for $\sin(x)$).

Solution: 0.9392

Explanation: The average (mean) value of a function is given by Equation 23.5. Therefore, we can simply substitute this in to the equation:

$$\text{average} = \frac{\int_1^{1.5} \sin(x)dx}{1.5 - 1}$$
$$= \frac{\int \sin(x)\,dx \Big|_1^{1.5}}{1.5 - 1}$$
$$= \frac{-\cos(x) + C \Big|_1^{1.5}}{1.5 - 1}$$
$$= \frac{-\cos(1.5) - -\cos(1)}{1.5 - 1}$$
$$\approx \frac{-0.0707 - -0.5403}{1.5 - 1}$$
$$\approx \frac{0.4696}{0.5}$$
$$\approx 0.9392$$

6. **Question:** Bacterial populations can undergo rapid growth given the right conditions. The size of the population is represented by the variable p. The instantaneous growth rate of a population of E. coli bacteria is found to be 150% per hour. Using the general assumptions of calculus, find out an equation for the number of bacterial cells at any given time t where t is measured in hours. Use p_0 to represent the starting population at time $t = 0$.

Solution: $p = p_0\, e^{1.5t}$

Explanation: Since the growth rate is

given as a percentage of the existing population, we will start by converting that into a quantified rate based on the existing population. So, if the growth rate is 150% of the given population size per hour, then we can say that the instantaneous growth rate in terms of the number of cells per hour (instead of a percentage) can be given by $1.5p$. In other words, we can say that:

$$\frac{\mathrm{d}p}{\mathrm{d}t} = 1.5p$$

From this, we can determine the actual accumulation of p by rearranging the problem and integrating like this:

$$\frac{\mathrm{d}p}{\mathrm{d}t} = 1.5p$$

$$\frac{\mathrm{d}p}{p} = 1.5\mathrm{d}t$$

$$\int \frac{\mathrm{d}p}{p} = \int 1.5\mathrm{d}t$$

$$\ln(p) = 1.5t + C$$

We can solve for p by itself by take e and raising e to the power of each side:

$$\ln(p) = 1.5t + C$$

$$e^{\ln(p)} = e^{1.5t+C}$$

$$p = e^{1.5t+C}$$

$$p = e^{C}\, e^{1.5t}$$

Because C is just as arbitrary of a constant as e^{C}, we can reduce e^{C} to C and get:

$$p = C\, e^{1.5t}$$

The starting population at time $t = 0$ is p_0. Therefore, we can see that:

$$p = C\, e^{1.5t}\, p_0 \qquad = C\, e^{1.5(0)}$$

$$= C\, e^{0}$$

$$= C \cdot 1$$

$$= C$$

Therefore C represents our initial population p_0. So we can rewrite the equation as:

$$p = p_0\, e^{1.5t}$$

7. **Question:** Using the equation generated in the previous question, find out how long it would take for 100 cells to grow into $1,000,000$ cells.

Solution: The population of bacteria will reach $1,000,000$ cells in approximately 6.1402 hours.

Explanation: The equation from the previous problem is:

$$p = p_0\, e^{1.5t}$$

Since we start with 100 cells, that is our p_0. Therefore, the equation becomes:

$$p = 100\, e^{1.5t}$$

Now, the goal is to solve for t when the population is $1,000,000$. Therefore, we just plug in and solve:

$$p = 100\, e^{1.5t}$$

$$1,000,000 = 100\, e^{1.5t}$$

$$\frac{1,000,000}{100} = e^{1.5t}$$

$$10,000 = e^{1.5t}$$

$$\ln(10,000) = \ln(e^{1.5t})$$

$$9.2103 \approx 1.5t$$

$$\frac{9.2103}{1.5} \approx t$$

$$6.1402 \approx t$$

The population of bacteria will reach $1,000,000$ cells in approximately 6.1402 hours.

8. **Question:** Find the average value of the function $x^3 + 12x^2 - 5$ from $x = 6$ to $x = 8$.

Solution: the average value on this interval is 937.

Explanation: The average (mean) value of a function is given by Equation 23.5. We can use this to solve the problem directly:

$$\text{average} = \frac{\int_6^8 (x^3 + 12x^2 - 5)\, \mathrm{d}x}{8 - 6}$$

$$= \frac{\int (x^3 + 12x^2 - 5)\, \mathrm{d}x \Big|_6^8}{8 - 6}$$

$$= \frac{\frac{x^4}{4} + 4x^3 - 5x + C \Big|_6^8}{8 - 6}$$

$$= \frac{\frac{(8)^4}{4} + 4(8)^3 - 5(8) - \frac{(6)^4}{4} - 4(6)^3 - 5(6)}{8 - 6}$$

$$= \frac{1024 + 2048 - 40 - 324 - 864 + 30}{8 - 6}$$

$$= \frac{1874}{2}$$

$$= 937$$

9. **Question:** In a probability distribution, what is the *total* area under the entire probability distribution?

Solution: 1

Explanation: Since area under a distribution curve represents the probability that some particular value of x will occur, the probability that *any* value of x will occur is 100% (assuming, of course, that whatever it is that you are taking the probability of happens at all). When dealing with probabilities as decimals instead of percents, 100% is represented as 1.

Therefore, since the probability that x will attain any value whatsoever is 1, and the area under the probability distribution curve represents the probability that x will attain a value within that range, the total area under the probability distribution curve will be 1.

10. **Question:** If someone told you the following probabilities (given as a decimal) for the heights of a type of organism, what would you think was wrong with those probabilities?

- < 30cm: 0.4
- 30cm–60cm: 0.5
- > 60cm: 0.3

Solution: The probabilities add up to more than 1.

11. **Question:** Radium-226 decays at an instantaneous rate of $\frac{0.0436\%}{\text{year}}$. Using the variable r for the amount of Radium-226 in grams, find the equation to represent the amount of Radium-226 left after time t, where t is in years, and r_0 is the starting amount of Radium-226.

Solution: $r = r_0\, e^{-0.000436t}$

Explanation: This is exactly like compound accumulation, except that the rate is negative. First, we need to convert the percent value into a decimal, so $0.0436\% = 0.00436$. Therefore, the decay rate in grams of Radium-226 is given by $\frac{\mathrm{d}r}{\mathrm{d}t}$, where $\frac{\mathrm{d}r}{\mathrm{d}t} = -0.000436r$. So, we can solve this for an equation for r in terms of t:

$$\frac{\mathrm{d}r}{\mathrm{d}t} = -0.000436r$$

$$\mathrm{d}r = -0.000436r\, \mathrm{d}t$$

$$\frac{\mathrm{d}r}{r} = -0.000436\, \mathrm{d}t$$

$$\int \frac{\mathrm{d}r}{r} = \int -0.000436\, \mathrm{d}t$$

$$\ln(r) = -0.000436t + C$$

$$e^{\ln(r)} = e^{-0.000436t + C}$$

$$r = e^{-0.000436t + C}$$

$$r = e^C e^{-0.000436t}$$

$$r = C e^{-0.000436t}$$

Remember, the last step works because e^C is just as arbitrary of a constant as C.

Now, to solve for C, we will set $t = 0$, which will mean that $r = r_0$. So, we can say:

$$r = C e^{-0.000436t}$$

$$r_0 = C e^{-0.000436(0)}$$

$$r_0 = C e^0$$

$$r_0 = C$$

So, the final equation will be:

$$r = r_0\, e^{-0.000436t}$$

12. **Question:** A pot of soup on the stove starts at 370K (Kelvin), and the ambient temperature of the room is 300K. Say that the cooling constant is $\frac{0.001}{\sec}$. Begin with Newton's Law of Cooling, and build an equation which tells you the temperature of the pot at time t (where t is in seconds).

Solution:

$$y = (70\text{K})e^{-\frac{0.001t}{\sec}} + 300\text{K}$$

Explanation: Newton's Law of Cooling states that, for a temperature of an object y, ambient temperature A, and cooling constant k, the change in the temperature of the object can be represented by:

$$\frac{dy}{dt} = -k(y - A)$$

In the question, the ambient temperature is 300K and the constant is $\frac{0.001}{\sec}$. Therefore, the equation becomes:

$$\frac{dy}{dt} = -\frac{0.001}{\sec}(y - 300\text{K})$$

We can use u-substitution, where $u = y - 300\text{K}$, and $du = dy$, which yields:

$$\frac{du}{dt} = -\frac{0.001}{\sec}u$$

Now we multiply by $\frac{dt}{u}$ in order to move the variables to their own sides:

$$\frac{du}{u} = -\frac{0.001}{\sec} dt$$

Now we can integrate:

$$\int \frac{du}{u} = \int -\frac{0.001}{\sec} dt$$

$$\ln(u) = -\frac{0.001t}{\sec} + C$$

Now we exponentiate both sides to get rid of the natural log:

$$e^{\ln(u)} = e^{-\frac{0.001t}{\sec} + C}$$

$$u = Ce^{-\frac{0.001t}{\sec}}$$

Now we can back-substitute for u:

$$y - 300\text{K} = Ce^{-\frac{0.001t}{\sec}}$$

$$y = Ce^{-\frac{0.001t}{\sec}} + 300\text{K}$$

Now we can solve for C when $t = 0$:

$$y_0 = Ce^{-\frac{0.001(0)}{\sec}} + 300\text{K}$$

$$y_0 = Ce^0 + 300\text{K}$$

$$y_0 = C + 300\text{K}$$

$$C = y_0 - 300\text{K}$$

Now, substituting C back into the equation, we get:

$$y = (y_0 - 300\text{K})e^{-\frac{0.001t}{\sec}} + 300\text{K}$$

However, the problem actually gives the value for y_0 as 370K, so we can substitute that into the problem, yielding:

$$y = (370\text{K} - 300\text{K})e^{-\frac{0.001t}{\sec}} + 300\text{K}$$

$$y = (70\text{K})e^{-\frac{0.001t}{\sec}} + 300\text{K}$$

13. **Question:** Given the equation from the previous question, what is the temperature of the pot after 500 seconds?

Solution: 342.4550K

Explanation: The equation from the previous question was:

$$y = (70\text{K})e^{-\frac{0.001t}{\sec}} + 300\text{K}$$

To find the temperature after 500 seconds, just set t to be 500sec and solve:

$$y = (70\text{K})e^{-\frac{0.001(500\text{sec})}{\sec}} + 300\text{K}$$

$$= (70\text{K})e^{-0.5} + 300\text{K}$$

$$\approx (70\text{K})(0.6065) + 300\text{K}$$

$$\approx 42.4550\text{K} + 300\text{K}$$

$$\approx 342.4550\text{K}$$

Chapter 24

Numeric Integration Techniques

Find the following definite integrals using a Left Riemann Sum:

1. **Question:** Find $\int_1^2 x^2 \, dx$ using $n = 4$ partitions.

   Solution: 1.96875

   Explanation: We start by finding Δx using Equation 24.1:

   $$\Delta x = \frac{2 - 1}{4} = 0.25$$

   Next, we create the summation equation using Equation 24.4:

   $$\int_1^2 f(x) \, dx \approx \sum_{k=1}^{4} f(1 + (k - 1) \cdot 0.25) \cdot 0.25$$

   Next, we need to create a table of the values that we will use:

k	Δx	x_k	$f(x_k)$	area
1	0.25	1	1	0.25
2	0.25	1.25	1.5625	0.390625
3	0.25	1.5	2.25	0.5625
4	0.25	1.75	3.0625	0.765625

   Summing all of the values up will yield our answer:

   $$0.25 + 0.390625 + 0.5625 + 0.765625 = 1.96875$$

2. **Question:** Find $\int_0^3 e^x \, dx$ using $n = 3$ partitions.

   Solution: 11.1074

   Explanation: We start by finding Δx using Equation 24.1:

   $$\Delta x = \frac{3 - 0}{3} = 1$$

   Next, we create the summation equation using Equation 24.4:

   $$\int_0^3 f(x) \, dx \approx \sum_{k=1}^{3} f(0 + (k - 1) \cdot 1) \cdot 1$$

   This simplifies to:

   $$\int_0^3 f(x) \, dx \approx \sum_{k=1}^{3} f(k - 1)$$

   Next, we need to create a table of the values that we will use:

k	Δx	x_k	$f(x_k)$	area
1	1	0	1	1
2	2	1	2.7183	2.7183
3	3	2	7.3891	7.3891

   Summing all of the values up will yield our answer:

   $$1 + 2.7183 + 7.3891 = 11.1074$$

3. **Question:** Find $\int_0^2 \sin(x^2) \, dx$ using $n = 4$ partitions.

   Solution: 1.1592

   Explanation: We start by finding Δx using Equation 24.1:

   $$\Delta x = \frac{2 - 0}{4} = 0.5$$

Next, we create the summation equation using Equation 24.4:

$$\int_a^b f(x)\,\mathrm{d}x \approx \sum_{k=1}^{4} f(0 + (k-1) \cdot 0.5) \cdot 0.5$$

This simplifies to:

$$\int_a^b f(x)\,\mathrm{d}x \approx \sum_{k=1}^{4} f(0.5k - 0.5) \cdot 0.5$$

Next, we need to create a table of the values that we will use:

k	Δx	x_k	$f(x_k)$	area
1	0.5	0	0	0
2	0.5	0.5	0.4794	0.2397
3	0.5	1	0.8415	0.42075
4	0.5	1.5	0.9975	0.49875

Summing all of the values up will yield our answer:

$$0 + 0.2397 + 0.42075 + 0.49875 = 1.1592$$

Find the following numeric integrals using a Right Riemann Sum:

4. **Question:** Find $\int_2^3 \ln(x)\,\mathrm{d}x$ using $n = 3$ partitions.

Solution: 0.8403

Explanation: We start by finding Δx using Equation 24.1:

$$\Delta x = \frac{2 - 3}{3} = \frac{1}{3}$$

Next, we create the summation equation using Equation 24.4:

$$\int_2^3 f(x)\,\mathrm{d}x \approx \sum_{k=1}^{3} f(2 + k \cdot \frac{1}{3}) \cdot \frac{1}{3}$$

Next, we need to create a table of the values that we will use:

k	Δx	x_k	$f(x_k)$	area
1	$\frac{1}{3}$	2	0.6931	0.2310
2	$\frac{1}{3}$	2.3333	0.8473	0.2824
3	$\frac{1}{3}$	2.6667	0.9808	0.3269

Summing all of the values up will yield our answer:

$$0.2310 + 0.2824 + 0.3269 = 0.8403$$

5. **Question:** Find $\int_5^7 x \cos(x)\,\mathrm{d}x$ using $n = 4$ partitions.

Solution: 8.7124

Explanation: We start by finding Δx using Equation 24.1:

$$\Delta x = \frac{5 - 7}{4} = 0.5$$

Next, we create the summation equation using Equation 24.4:

$$\int_5^7 f(x)\,\mathrm{d}x \approx \sum_{k=1}^{4} f(5 + k \cdot 0.5) \cdot 0.5$$

Next, we need to create a table of the values that we will use:

k	Δx	x_k	$f(x_k)$	area
1	0.5	5	1.4183	0.70915
2	0.5	5.5	3.8977	1.94885
3	0.5	6	5.7610	2.8805
4	0.5	6.5	6.3478	3.1739

Summing all of the values up will yield our answer:

$$0.70915 + 1.94885 + 2.8805 + 3.1739 = 8.7124$$

Part IV

Manipulating Infinity

Chapter 25

Modeling Functions with Polynomial Series

1. **Question:** Use the process in the book in Sections 25.5 and 25.6 to find the first six terms for the polynomial expansion of $\sin(x)$.

   Solution: The coefficients are $0, 1, 0, -\frac{1}{6}, 0, \frac{1}{120}$. The terms of the expansion are therefore:

   $$\sin(x) \approx x + -\frac{1}{6}x^3 + \frac{1}{120}x^5$$

   Explanation: Let's begin by setting up the polynomial:

   $$\sin(x) = C_0 + C_1 x + C_2 x^2 + C_3 x^3 + C_4 x^4 + C_5 x^5 + \ldots$$

   Now, set $x = 0$ and we will see that all of the xs drop off:

   $$\sin(0) = C_0 + C_1 0 + C_2 0^2 + C_3 0^3 + C_4 0^4 + C_5 0^5 + \ldots$$

   $$\sin(0) = C_0 + 0 + 0 + 0 + 0 + 0$$

   $$\sin(0) = C_0$$

   We know that $\sin(0) = 0$, so that means that:

   $$\sin(0) = 0 = C_0$$

   So, $C_0 = 0$. That's the first coefficient. If we take the derivative of both sides of our polynomial equation, we get:

   $$\cos(x) = C_1 + 2C_2 x + 3C_3 x^2 + 4C_4 x^3 + 5C_5 x^4 + \ldots$$

   If we set $x = 0$ we get:

   $$\cos(0) = C_1 + 2C_2 0 + 3C_3 0^2 + 4C_4 0^3 + 5C_5 0^4 + \ldots$$

   $$\cos(0) = C_1 + 0 + 0 + 0 + 0$$

   $$\cos(0) = C_1$$

   We know that $\cos(0) = 1 = C_1$. So $C_1 = 1$. That's our second coefficient.

   To find the next coefficient, take the derivative of both sides again:

   $$-\sin(x) = 2C_2 + 6C_3 x + 12C_4 x^2 + 20C_5 x^3 + \ldots$$

   Setting $x = 0$ gives us:

   $$-\sin(0) = 2C_2 + 6C_3 0 + 12C_4 0^2 + 20C_5 0^3 + \ldots$$

   $$-\sin(0) = 2C_2 + 0 = 2C_2$$

   Since $-\sin(0) = 0$, then $2C_2 = 0$, so $C_2 = 0$. For the next coefficient, take the derivative of both sides again:

   $$-\cos(x) = 6C_3 + 24C_4 x + 60C_5 x^2 + \ldots$$

   Setting $x = 0$, we find:

   $$-\cos(0) = 6C_3 + 24C_4 0 + 60C_5 0^2 + \ldots$$

   $$-\cos(0) = 6C_3 + 0 = 6C_3$$

   Since we know that $-\cos(0) = -1$, we can say that $6C_3 = -1$, or that $C_3 = -\frac{1}{6}$.

   We perform the derivative again for the next coefficient:

   $$\sin(x) = 24C_4 + 120C_5 x + \ldots$$

   At $x = 0$, this becomes:

   $$\sin(0) = 24C_4 + 0 = 0$$

   So, $C_4 = 0$. We can take the derivative again for our last coefficient:

   $$\cos(x) = 120C_5 + \ldots$$

   At $x = 0$ this becomes:

   $$\cos(0) = 120C_5 + 0 = 1$$

So, $120C_5 = 1$, so $C_5 = \frac{1}{120}$.

Now we have done the first six coefficients! They were:

$$0, 1, 0, -\frac{1}{6}, 0, \frac{1}{120}$$

The equation is therefore:

$$\sin(x) \approx x + -\frac{1}{6}x^3 + \frac{1}{120}x^5$$

2. **Question:** Use the process in the book in Sections 25.5 and 25.6 to find the first six terms for the polynomial expansion of 2^x.

Solution: The   coefficients   are $1, 0.6931, 0.2402, 0.0555, 0.0096, 0.0013$.   The terms of the equation are:

$$2^x \approx 1 + 0.6931x + 0.2402x^2 + 0.0555x^3$$
$$+ 0.0096x^4 + 0.0013x^5$$

Explanation: Let's begin by setting up the polynomial:

$$2^x = C_0 + C_1x + C_2x^2 + C_3x^3 + C_4x^4 + C_5x^5 + \ldots$$

Now, set $x = 0$ and we will see all of the coefficients of x drop off:

$$2^0 = C_0 + C_10 + C_20^2 + C_30^3 + C_40^4 + C_50^5 + \ldots$$

$$2^0 = C_0 + 0 + 0 + 0 + 0 + 0 + \ldots$$

$$2^0 = C_0$$

Because of exponent rules, we know that $2^0 = 1$. Therefore,

$$2^0 = C_0 = 1$$

So the first coefficient is 1. For the next coefficient we need to take the derivative of both sides:

$$\ln(2)2^x = C_1 + 2C_2x + 3C_3x^2 + 4C_4x^3 + 5C_5x^4 + \ldots$$

At $x = 0$ this becomes:

$$\ln(2)2^0 = C_1 + 0 + 0 + 0 + 0 + \ldots$$

$$\ln(2)1 = C_1$$

$$\ln(2) = C_1$$

$$0.6931 \approx C_1$$

Therefore, the coefficient C_1 is $\ln(2)$ or 0.6931. For the next coefficient, we take the derivative again:

$$(\ln(2))^22^x = 2C_2 + 6C_3x + 12C_4x^2 + 20C_5x^3 + \ldots$$

Setting $x = 0$, we get:

$$(\ln(2))^22^0 = 2C_2 + 0 + 0 + 0 + \ldots$$

$$(\ln(2))^2 = 2C_2$$

$$C_2 = \frac{((\ln(2))^2}{2}$$

$$C_2 \approx 0.2402$$

Therefore, C_2 is approximately 0.2402. Taking the derivative again, we find:

$$(\ln(2))^32^x = 6C_3 + 24C_4x + 60C_5x^2 + \ldots$$

Setting $x = 0$, this becomes:

$$(\ln(2))^32^0 = 6C_3 + 0 + 0 + \ldots$$

$$(\ln(2))^3 = 6C_3$$

$$C_3 = \frac{(\ln(2))^3}{6}$$

$$C_3 \approx 0.0555$$

Taking the derivative again, we find:

$$(\ln(2))^42^x = 24C_4 + 120C_5x + \ldots$$

Setting $x = 0$ yields:

$$(\ln(2))^42^0 = 24C_4 + 0 + \ldots$$

$$(\ln(2))^4 = 24C_4$$

$$C_4 = \frac{(\ln(2))^4}{24}$$

$$C_4 \approx 0.0096$$

To find the next coefficient, take the derivative again:

$$(\ln(2))^52^x = 120C_5 + \ldots$$

Setting $x = 0$, we find:

$$(\ln(2))^52^0 = 120C_5 + \ldots$$

$$C_5 = \frac{(\ln(2))^5}{120}$$

$$C_5 \approx 0.0013$$

Therefore,   the   coefficients   are $1, 0.6931, 0.2402, 0.0555, 0.0096, 0.0013$.   That means that the equation is:

$$2^x = 1 + 0.6931x + 0.2402x^2 + 0.0555x^3$$
$$+ 0.0096x^4 + 0.0013x^5$$

3. **Question:** Using the previous two problems, find an approximate value for $\sin(1.1)$ and $2^{1.1}$. Compare these answers to what your calculator yields (don't forget to use radians!).

Solution: Using the equations, we find that $\sin(1.1) \approx 0.8916$ and $2^{1.1} \approx 2.1431$. Compared to a calculator, these are both approximately correct to three decimal places to the right of the decimal.

Explanation: The equation we found for $\sin(x)$ is:

$$\sin(x) \approx x + -\frac{1}{6}x^3 + \frac{1}{120}x^5$$

Substituting in 1.1 for x yields:

$$\sin(1.1) \approx 1.1 + -\frac{1}{6}(1.1)^3 + \frac{1}{120}(1.1)^5$$

$$\approx 1.1 + -\frac{1}{6}1.331 + \frac{1}{120}1.61051$$

$$\approx 1.1 - 0.2218 + 0.0134$$

$$\approx 0.8916$$

Using a calculator gives me approximately $0.891207\ldots$, which means that my result is correct for the first three digits after the decimal.

For calculating 2^x, the formula we discovered was:

$$2^x \approx 1 + 0.6931x + 0.2402x^2 + 0.0555x^3$$
$$+ 0.0096x^4 + 0.0013x^5$$

Using $x = 1.1$, we find:

$$2^{1.1} \approx 1 + 0.6931(1.1) + 0.2402(1.1)^2 + 0.0555(1.1)^3$$
$$+ 0.0096(1.1)^4 + 0.0013(1.1)^5$$
$$\approx 1 + 0.76241 + 0.290642 + 0.0738705$$
$$+ 0.01405536 + 0.002093663$$
$$\approx 2.143071523$$
$$\approx 2.1431$$

The calculator gives me $2^{1.1} \approx 2.1435469\ldots$. This means my answer is correct to the first three decimal places.

4. **Question:** Try to come up with a formula for $\sin(x)$ that is similar in style to Equation 25.9.

Solution:

$$\sin(x) = \sum_{n=0}^{\infty} \left((-1)^{\frac{n-1}{2}}\right)\left(\frac{\frac{1+(-1)^{n-1}}{2}}{n!}\right)x^n$$

Explanation: If you notice, the coefficients for $\sin(x)$ cycle exactly like for $\cos(x)$, but they are offset by one. $\cos(x)$ starts with 1 while $\sin(x)$ starts with zero. The equation for $\cos(x)$ was given as:

$$\cos(x) = \sum_{n=0}^{\infty} \left((-1)^{\frac{n}{2}}\right)\left(\frac{\frac{1+(-1)^{n}}{2}}{n!}\right)x^n$$

To get $\sin(x)$ to behave similarly, we need to just offset n by one:

$$\sin(x) = \sum_{n=0}^{\infty} \left((-1)^{\frac{n-1}{2}}\right)\left(\frac{\frac{1+(-1)^{n-1}}{2}}{n!}\right)x^n$$

That is the answer.

5. **Question:** Estimate the value of e using the first ten terms of the expansion of e^x. Round off to eight decimal places. Look up the value of e in a book or on the web. How many decimal places were you accurate to?

Solution: The estimate is 2.71828153 and it is accurate in its first six decimal places.

Explanation: The value of e can be evaluated using the polynomial expansion of e^x, and then setting $x = 1$. The polynomial expansion of e^x is given by Equation 25.11 as:

$$e^x = \sum_{n=0}^{\infty} \frac{1}{n!}x^n$$

Now, since we are just looking for e, that means that we are looking for e^1, which means that x simply becomes 1. Because 1 raised to any power is just 1, then that means that all powers of x can be dropped from the equation, because they will just become identity multipliers.

Therefore, the equation for e by itself is:

$$e = \sum_{n=0}^{\infty} \frac{1}{n!}$$

The first ten terms of this will be from $n = 0$ to $n = 9$. This will be:

$$\frac{1}{0!} + \frac{1}{1!} + \frac{1}{2!} + \frac{1}{3!} + \frac{1}{4!} + \frac{1}{5!} + \frac{1}{6!} + \frac{1}{7!} + \frac{1}{8!} + \frac{1}{9!}$$

Now, we have discussed before that 0! is 1. Therefore, if we expand the factorials of all of these, we get:

$$\frac{1}{1} + \frac{1}{1} + \frac{1}{2} + \frac{1}{6} + \frac{1}{24} + \frac{1}{120} + \frac{1}{720} + \frac{1}{5040} + \frac{1}{40320} + \frac{1}{362880}$$

Calculating these, rounded off to eight decimal places, we have:

$$1 + 1 + 0.5 + 0.16666667 + 0.04166667$$

$$+ \, 0.00833333 + 0.00138889 + 0.00019841$$

$$+ \, 0.00002480 + 0.00000276$$

Adding these together yields 2.71828153.

Looking up the value of e to ten decimal places yields 2.7182818284... (and, actually, that last four would round up to 5). Comparing to our result shows that our system is accurate in the first six decimal places.

6. **Question:** Use Equation 25.8 to find the $n = 8$ term for the expansion of e^x.

Solution: $\frac{1}{40320} x^8$

Explanation: The general formula given by Equation 25.8 is:

$$f(x) = \sum_{n=0}^{\infty} \frac{f^n(0)}{n!} x^n$$

In this formula, f^n represents the nth derivative of f (in this case f is e^x). Therefore, the $n = 8$ term will be:

$$\frac{f^8(0)}{8!} x^8$$

In this formula, $f^8(0)$ represents the eighth derivative of e^x at zero. Every derivative of e^x is just e^x, so this is e^x evaluated at $x = 0$, or just e^0, which is 1. Therefore, it reduces to:

$$\frac{1}{8!} x^8$$

8! is 40320, so this becomes:

$$\frac{1}{40320} x^8$$

7. **Question:** Using Equation 25.8, find the first four terms of a polynomial representation of $x \, e^x$.

Solution:

$$x \, e^x \approx x + x^2 + \frac{3}{6} x^3$$

There are only three terms listed because the first term is zero.

Explanation: Equation 25.8 says:

$$f(x) = \sum_{n=0}^{\infty} \frac{f^n(0)}{n!} x^n$$

Let's look at the first four terms of this expansion:

$$x \, e^x \approx \frac{f(0)}{0!} x^0 + \frac{f'(0)}{1!} x^1 + \frac{f''(0)}{2!} x^2 + \frac{f'''(0)}{3!} x^3$$

This simplifies to:

$$x \, e^x \approx f(0) + f'(0) \, x + \frac{f''(0)}{2} x^2 + \frac{f'''(0)}{6} x^3$$

To complete the first four terms of this expansion, we are going to need the first three derivatives of the function. Then we will need to evaluate these at $x = 0$. These are:

$$f(x) = x \, e^x$$
$$f(0) = 0$$
$$f'(x) = x \, e^x + e^x$$
$$f'(0) = 0 + e^0 = 1$$
$$f''(x) = x \, e^x + e^x + e^x$$
$$f''(0) = 0 + e^0 + e^0 = 2$$
$$f'''(x) = x \, e^x + e^x + e^x + e^x$$
$$f'''(0) = 0 + e^0 + e^0 + e^0 = 3$$

Plugging these into our formula yields:

$$x \, e^x \approx 0 + 1x + \frac{2}{2} x^2 + \frac{3}{6} x^3$$

This simplifies to:

$$x \, e^x \approx x + x^2 + \frac{3}{6} x^3$$

8. **Question:** Use the previous question to write a formula for every term of the expansion of $x\,e^x$.

Solution:

$$x\,e^x = \sum_{n=0}^{\infty} \frac{1}{(n-1)!} x^n$$

Explanation: If you notice from above, every derivative at zero just goes up by one. Therefore, $f^n(0)$ is always n. This means that our formula can be written as:

$$x\,e^x = \sum_{n=0}^{\infty} \frac{n}{n!} x^n$$

We can write it even more simply when you realize that $\frac{n}{n!}$ is the same as $\frac{1}{(n-1)!}$. Then the formula becomes:

$$x\,e^x = \sum_{n=0}^{\infty} \frac{1}{(n-1)!} x^n$$

9. **Question:** Write out the expansion of $\cos(x)$ to ten terms (including the zeroed out terms) using Equation 25.9.

Solution:

$$\cos(x) \approx 1 + -\frac{1}{2}x^2 + \frac{1}{24}x^4 + -\frac{1}{720}x^6 + \frac{1}{40320}x^8 +$$

Explanation: Equation 25.9 says:

$$\cos(x) = \sum_{n=0}^{\infty} \left((-1)^{\frac{n}{2}}\right)\left(\frac{\frac{1+(-1)^n}{2}}{n!}\right) x^n$$

If we write out the first ten terms, we get:

$$\cos(x) \approx \left((-1)^{\frac{0}{2}}\right)\left(\frac{\frac{1+(-1)^0}{2}}{0!}\right)x^0 + \left((-1)^{\frac{1}{2}}\right)\left(\frac{\frac{1+(-1)^1}{2}}{1!}\right)x^1$$

$$+ \left((-1)^{\frac{2}{2}}\right)\left(\frac{\frac{1+(-1)^2}{2}}{2!}\right)x^2 + \left((-1)^{\frac{3}{2}}\right)\left(\frac{\frac{1+(-1)^3}{2}}{3!}\right)x^3$$

$$+ \left((-1)^{\frac{4}{2}}\right)\left(\frac{\frac{1+(-1)^4}{2}}{4!}\right)x^4 + \left((-1)^{\frac{5}{2}}\right)\left(\frac{\frac{1+(-1)^5}{2}}{5!}\right)x^5$$

$$+ \left((-1)^{\frac{6}{2}}\right)\left(\frac{\frac{1+(-1)^6}{2}}{6!}\right)x^6 + \left((-1)^{\frac{7}{2}}\right)\left(\frac{\frac{1+(-1)^7}{2}}{7!}\right)x^7$$

$$+ \left((-1)^{\frac{8}{2}}\right)\left(\frac{\frac{1+(-1)^8}{2}}{8!}\right)x^8 + \left((-1)^{\frac{9}{2}}\right)\left(\frac{\frac{1+(-1)^9}{2}}{9!}\right)x^9$$

This simplifies to:

$$\cos(x) \approx 1 + 0 + -\frac{1}{2}x^2 + 0 + \frac{1}{24}x^4 + 0$$

$$+ -\frac{1}{720}x^6 + 0 + \frac{1}{40320}x^8 + 0$$

We can remove the zeroes, which leaves:

$$\cos(x) \approx 1 + -\frac{1}{2}x^2 + \frac{1}{24}x^4 + -\frac{1}{720}x^6 + \frac{1}{40320}x^8$$

10. **Question:** What is the term for $n = 513$ (i.e., the 514th term) of the expansion of $\cos(x)$ according to Equation 25.9? If there are any factorials in the result, leave them as factorials.

Solution: 0

Explanation: The formula is:

$$\cos(x) = \sum_{n=0}^{\infty} \left((-1)^{\frac{n}{2}}\right)\left(\frac{\frac{1+(-1)^n}{2}}{n!}\right) x^n$$

The piece that is important is $1 + (-1)^n$. $(-1)^{513}$ is -1. Therefore, this expression reduces to zero. A zero multiplier zeroes out the whole term. Therefore, the term is simply 0.

11. **Question:** What is the term for $n = 8214$ (i.e., the 8215th term) of the expansion of $\cos(x)$ according to Equation 25.9? If there are any

factorials in the result, leave them as factorials.

Solution: $\frac{-1}{8214!}x^{8214}$

Explanation: The formula is:

$$\cos(x) = \sum_{n=0}^{\infty} \left((-1)^{\frac{n}{2}}\right)\left(\frac{\frac{1+(-1)^n}{2}}{n!}\right)x^n$$

If we substitute 8214 in for n, we get:

$$\left((-1)^{\frac{8214}{2}}\right)\left(\frac{\frac{1+(-1)^{8214}}{2}}{8214!}\right)x^{8214}$$

$$\left((-1)^{4107}\right)\left(\frac{\frac{1+1}{2}}{8214!}\right)x^{8214}$$

$$(-1)\frac{1}{8214!}x^{8214}$$

$$\frac{-1}{8214!}x^{8214}$$

Chapter 26

Advanced Topics in Polynomial Series

1. **Question:** Use the first four terms of the Taylor expansion of $\ln(x)$ given in Equation 26.1 (starting with $n = 1$) to approximate the value of $\ln(0.5)$. Use a calculator to check to see how close the approximation is. Note that you should keep it in fractional form until the very end to remove as much approximation error as possible.

   Solution: The result using the series is -0.6615. The result using the calculator is -0.6931.

   Explanation: The series given in Equation 26.1 is

   $$\ln(x) = \sum_{n=1}^{\infty} (-1)^{n+1} \frac{1}{n} (x-1)^n$$

   The first four terms, then, are:

   $$\ln(x) \approx \frac{(-1)^2}{1}(x-1) + \frac{(-1)^3}{2}(x-1)^2$$
   $$+ \frac{(-1)^4}{3}(x-1)^3 + \frac{(-1)^5}{4}(x-1)^4$$

   This simplifies to:

   $$\ln(x) \approx (x-1) - \frac{1}{2}(x-1)^2 + \frac{1}{3}(x-1)^3 - \frac{1}{4}(x-1)^4$$

   Therefore, $\ln(0.5)$ can be calculated as follows:

   $$\ln(0.5) \approx (0.5-1) - \frac{1}{2}(0.5-1)^2$$
   $$+ \frac{1}{3}(0.5-1)^3 - \frac{1}{4}(0.5-1)^4$$

   $$\ln(0.5) \approx -0.5 - \frac{1}{2}(-0.5)^2 + \frac{1}{3}(-0.5)^3 - \frac{1}{4}(-0.5)^4$$

   $$\ln(0.5) \approx -0.5 - \frac{0.25}{2} + \frac{-0.125}{3} - \frac{0.0625}{4}$$

   $$\ln(0.5) \approx \frac{-6 - 1.5 - 0.25 - 0.1875}{12}$$

   $$\ln(0.5) \approx \frac{-7.9375}{12}$$

   $$\ln(0.5) \approx -0.6615$$

   The calculator, when calculating $\ln(0.5)$, yields -0.6931. Using just the first four terms made us off by a little more the 0.03.

2. **Question:** Use the first four terms of the Taylor expansion of $\ln(x)$ given in Equation 26.1 (starting with $n = 1$) to approximate the value of $\ln(1.3)$. Use a calculator to check to see how close the approximation is.

   Solution: The expansion yields a result of 0.2695. The calculator yields 0.2624.

   Explanation: The series given in Equation 26.1 is

   $$\ln(x) = \sum_{n=1}^{\infty} (-1)^{n+1} \frac{1}{n} (x-1)^n$$

   The first four terms, then, simplify to:

   $$\ln(x) \approx (x-1) - \frac{1}{2}(x-1)^2 + \frac{1}{3}(x-1)^3 - \frac{1}{4}(x-1)^4$$

Therefore, ln(1.3) can be calculated as follows:

$$\ln(1.3) \approx (1.3 - 1) - \frac{1}{2}(1.3 - 1)^2$$

$$+ \frac{1}{3}(1.3 - 1)^3 - \frac{1}{4}(1.3 - 1)^4$$

$$\ln(1.3) \approx 0.3 - \frac{1}{2}(0.3)^2 + \frac{1}{3}(0.3)^3 - \frac{1}{4}(0.3)^4$$

$$\ln(1.3) \approx 0.3 - \frac{0.09}{2} + \frac{0.027}{3} - \frac{0.0081}{4}$$

$$\ln(1.3) \approx \frac{3.6 - 0.45 + 0.108 - 0.0243}{12}$$

$$\ln(1.3) \approx \frac{3.2337}{12}$$

$$\ln(1.3) \approx 0.2695$$

The calculator gives ln(1.3) as 0.2624, so our result was pretty accurate.

3. **Question:** Use the first four terms of the Taylor expansion of $\ln(x)$ given in Equation 26.1 (starting with $n = 1$) to approximate the value of $\ln(1 - \frac{i}{2})$, where i is the imaginary unit.

Solution: The solution is approximately $0.1406 - 0.4583i$.

Explanation: The expansion we have been working with is:

$$\ln(x) \approx (x - 1) - \frac{1}{2}(x - 1)^2 + \frac{1}{3}(x - 1)^3 - \frac{1}{4}(x - 1)^4$$

Because this is just a polynomial, we can use $1 - \frac{i}{2}$ just as straightforwardly as any other value. So, substituting it in gives us:

$$\ln(1 - \frac{i}{2}) \approx (1 - \frac{i}{2} - 1) - \frac{1}{2}(1 - \frac{i}{2} - 1)^2$$

$$+ \frac{1}{3}(1 - \frac{i}{2} - 1)^3 - \frac{1}{4}(1 - \frac{i}{2} - 1)^4$$

$$\ln(1 - \frac{i}{2}) \approx (-\frac{i}{2}) - \frac{1}{2}(-\frac{i}{2})^2 + \frac{1}{3}(-\frac{i}{2})^3 - \frac{1}{4}(-\frac{i}{2})^4$$

$$\ln(1 - \frac{i}{2}) \approx -\frac{i}{2} - \frac{1}{2}\frac{-1}{4} + \frac{1}{3}\frac{i}{8} - \frac{1}{4}\frac{-1}{16}$$

$$\ln(1 - \frac{i}{2}) \approx -\frac{i}{2} + \frac{1}{8} + \frac{i}{24} - \frac{-1}{64}$$

$$\ln(1 - \frac{i}{2}) \approx \frac{9}{64} - \frac{11i}{24}$$

$$\ln(1 - \frac{i}{2}) \approx 0.1406 - 0.4583i$$

Question: Find the first six terms for the Taylor expansion of $\sin(x)$ about π. Zero-terms can be included in your count, so do $n = 0$ to $n = 5$.

Solution: The terms from $n = 0$ to $n = 5$ of the Taylor expansion of $\sin(x)$ about π are:

$$\sin(x) = -(x - \pi) + \frac{1}{6}(x - \pi)^3 + \frac{-1}{120}(x - \pi)^5$$

Explanation: The formula for a Taylor expansion about a number is given by Equation 26.2:

$$f(x) = \sum_{n=0}^{\infty} \frac{f^{(n)}(a)}{n!}(x - a)^n$$

Since we are doing the expansion about π, the first six terms (from $n = 0$ to $n = 5$) will be:

$$f(x) = \frac{f(\pi)}{0!}(x - \pi)^0 + \frac{f^{(1)}}{1!}(\pi)(x - \pi)$$

$$+ \frac{f^{(2)}}{2}(\pi)(x - \pi)^2 + \frac{f^{(3)}}{3!}(\pi)(x - \pi)^3$$

$$+ \frac{f^{(4)}}{4!}(\pi)(x - \pi)^4 + \frac{f^{(5)}}{5!}(\pi)(x - \pi)^5$$

Since f is sin, that means that $f^{(1)}$ is cos, f^2 is $-\sin$, etc. Additionally, the first few terms can be cleaned up quite a bit, and we can compute all of the factorials. Therefore, this becomes:

$$\sin(x) = \sin(\pi) + \cos(\pi)(x - \pi)$$

$$+ \frac{-\sin(\pi)}{2}(x - \pi)^2 + \frac{-\cos(\pi)}{6}(x - \pi)^3$$

$$+ \frac{\sin(\pi)}{24}(x - \pi)^4 + \frac{\cos(\pi)}{120}(x - \pi)^5$$

Now, since $\sin(\pi)$ is 0 and $\cos(\pi)$ is -1, this simplifies to:

$$\sin(x) = 0 + -(x - \pi) + 0 + \frac{1}{6}(x - \pi)^3 + 0 + \frac{-1}{120}(x - \pi)^5$$

$$\sin(x) = -(x - \pi) + \frac{1}{6}(x - \pi)^3 + \frac{-1}{120}(x - \pi)^5$$

4. **Question:** Find the first six terms for the Taylor expansion of $\cos(x)$ about π. Zero

terms are included in the count, so perform the expansion from $n = 0$ to $n = 5$.

Solution: The terms from $n = 0$ to $n = 5$ of the Taylor expansion of $\cos(x)$ about π are:

$$\cos(x) = -1 + \frac{1}{2}(x - \pi)^2 + \frac{-1}{24}(x - \pi)^4$$

Explanation: The formula for a Taylor expansion about a number is given by Equation 26.2:

$$f(x) = \sum_{n=0}^{\infty} \frac{f^{(n)}(a)}{n!}(x - a)^n$$

Since we are doing the expansion about π, the first six terms (from $n = 0$ to $n = 5$) will be:

$$f(x) = \frac{f(\pi)}{0!}(x - \pi)^0 + \frac{f^{(1)}}{1!}(\pi)(x - \pi)$$

$$+ \frac{f^{(2)}}{2!}(\pi)(x - \pi)^2 + \frac{f^{(3)}}{3!}(\pi)(x - \pi)^3$$

$$+ \frac{f^{(4)}}{4!}(\pi)(x - \pi)^4 + \frac{f^{(5)}}{5!}(\pi)(x - \pi)^5$$

Since f is cos, that means that $f^{(1)}$ is $-\sin$, f^2 is $-\cos$, etc. Additionally, the first few terms can be cleaned up quite a bit, and we can compute all of the factorials. Therefore, this becomes:

$$\cos(x) = \cos(\pi) + -\sin(\pi)(x - \pi) + \frac{-\cos(\pi)}{2}(x - \pi)^2$$

$$+ \frac{\sin(\pi)}{6}(x - \pi)^3 + \frac{\cos(\pi)}{24}(x - \pi)^4$$

$$+ \frac{-\sin(\pi)}{120}(x - \pi)^5$$

Now, since $\sin(\pi)$ is 0 and $\cos(\pi)$ is -1, this simplifies to:

$$\cos(x) = -1 + 0 + \frac{1}{2}(x - \pi)^2 + 0 + \frac{-1}{24}(x - \pi)^4 + 0$$

$$\cos(x) = -1 + \frac{1}{2}(x - \pi)^2 + \frac{-1}{24}(x - \pi)^4$$

5. **Question:** Use the Taylor expansion of $\cos(x)$ about π above to approximate the value of $\cos(4)$. Use 3.1416 for π.

Solution: $\cos(4)$ is approximately -0.6542.

Explanation: The first terms of the formula for $\cos(x)$ about π was determined to be:

$$\cos(x) = -1 + \frac{1}{2}(x - \pi)^2 + \frac{-1}{24}(x - \pi)^4$$

Therefore, taking π to be 3.1416, $\cos(4)$ will be:

$$\cos(4) = -1 + \frac{1}{2}(4 - 3.1416)^2 + \frac{-1}{24}(4 - 3.1416)^4$$

$$\cos(4) = -1 + \frac{1}{2}(0.8584)^2 + \frac{-1}{24}(0.8584)^4$$

$$\cos(4) \approx -1 + \frac{1}{2}(0.73685) + \frac{-1}{24}(0.5429487)$$

$$\cos(4) \approx -0.6542$$

6. **Question:** Use the Taylor expansion of $\cos(x)$ about π above to approximate the value of $\cos(3 + 2i)$. Use 3.1416 as an approximation of π

Solution:

$$\cos(3 + 2i) \approx -3.6366 - 0.4711i$$

Explanation: The expansion of $\cos(x)$ about π is:

$$\cos(x) \approx -1 + \frac{1}{2}(x - \pi)^2 + \frac{-1}{24}(x - \pi)^4$$

Substituting in 3.1416 for π and $3 + 2i$ for x yields the following:

$$\cos(3 + 2i) \approx -1 + \frac{1}{2}(3 + 2i - 3.1416)^2$$

$$+ \frac{-1}{24}(3 + 2i - 3.1416)^4$$

$$\cos(3 + 2i) \approx -1 + \frac{1}{2}(-3.97995 - 0.5664i)$$

$$+ \frac{-1}{24}(15.5192 + 4.50849i)$$

$$\cos(3 + 2i) \approx -3.6366 - 0.4711i$$

7. **Question:** Find the first four terms of the series expansion of $\log_{10}(x)$ (i.e., the base 10

logarithm of x) about 100. You will need to approximate decimals to eight places. Can you determine where this will be valid for?

Solution:

$$\log_{10}(x) \approx 2 + 0.00434294(x - 100)$$
$$+ \; -0.00002172(x - 100)^2$$
$$+ \; 0.00000015(x - 100)^3$$

This solution is valid for $|x - 100| < 1$.

Explanation: To find a series expansion for $\log_{10}(x)$ about 100, we will start with the general formula for a Taylor series (Equation 26.2):

$$f(x) = \sum_{n=0}^{\infty} \frac{f^{(n)}(a)}{n!}(x - a)^n$$

Using $a = 100$, we can expand this to the first four terms as follows:

$$\log_{10}(x) \approx \frac{\log_{10}(100)}{0!}(x - 100)^0$$
$$+ \frac{\log_{10}'(100)}{1!}(x - 100)^1$$
$$+ \frac{\log_{10}''(100)}{2!}(x - 100)^2$$
$$+ \frac{\log_{10}'''(100)}{3!}(x - 100)^3$$

$$\log_{10}(x) \approx \log_{10}(100) + \log_{10}'(100)(x - 100)$$
$$+ \frac{\log_{10}''(100)}{2}(x - 100)^2$$
$$+ \frac{\log_{10}'''(100)}{6}(x - 100)^3$$

We can solve for the derivatives as follows:

$$f(x) = \log_{10}(x)$$
$$f(100) = \log_{10}(100) = 2$$
$$f'(x) = \frac{1}{x \ln(10)}$$
$$f'(100) = \frac{1}{100 \ln(10)} = 0.00434294$$
$$f''(x) = -\frac{1}{x^2 \ln(10)}$$
$$f''(100) = -\frac{1}{100^2 \ln(10)} = -0.00004343$$
$$f'''(x) = \frac{2}{x^3 \ln(10)}$$
$$f'''(100) = \frac{2}{100^3 \ln(10)} = 0.00000087$$

Therefore, the expansion becomes:

$$\log_{10}(x) \approx 2 + 0.00434294(x - 100)$$
$$+ \frac{-0.00004343}{2}(x - 100)^2$$
$$+ \frac{0.00000087}{6}(x - 100)^3$$

$$\log_{10}(x) \approx 2 + 0.00434294(x - 100)$$
$$+ \; -0.00002172(x - 100)^2$$
$$+ \; 0.00000015(x - 100)^3$$

8. **Question:** Use the expansion of $\log_{10}(x)$ about 100 to estimate $\log_{10}(95)$.

Solution: $\log_{10}(95) \approx 1.9778$

Explanation: The expansion we deduced from the previous question is:

$$\log_{10}(x) \approx 2 + 0.00434294(x - 100)$$
$$+ \; -0.00002172(x - 100)^2$$
$$+ \; 0.00000015(x - 100)^3$$

Therefore, we can solve for $\log_{10}(95)$ as follows:

$$\log_{10}(95) \approx 2 + 0.00434294(95 - 100)$$
$$+ \; -0.00002172(95 - 100)^2$$
$$+ \; 0.00000015(95 - 100)^3$$

$$\log_{10}(95) \approx 2 + 0.00434294(-5)$$
$$+ \; -0.00002172(-5)^2$$
$$+ \; 0.00000015(-5)^3$$

$$\log_{10}(95) \approx 2 + 0.00434294(-5)$$
$$+ \; -0.00002172(25)$$
$$+ \; 0.00000015(125)$$

$$\log_{10}(95) \approx 1.9778$$

(a) **Question:** What are the two main ways to integrate a non-integratable function?

Solution: The two main ways are to

(a) simply assign a name to a function that represents the integral result, and (b) to convert the function to an infinite series, and then integrate the series term.

(b) **Question:** Find the series representation for the solution of $\int \cos(x^3)\, dx$.

Solution:

$$\int \cos(x^3)\, dx = \sum_{n=0}^{\infty} ((-1)^n) \left(\frac{1}{2n!}\right) \frac{x^{6n+1}}{6n+1}$$

Explanation: Since we know the series representation for $\cos(x)$, we can replace x with x^3 in the expansion to get the proper result. If we replace x with x^3 in Equation 25.10, we will get:

$$\cos(x^3) = \sum_{n=0}^{\infty} ((-1)^n) \left(\frac{1}{2n!}\right) x^{6n}$$

If we integrate each term with the power rule, we get:

$$\sum_{n=0}^{\infty} ((-1)^n) \left(\frac{1}{2n!}\right) \frac{x^{6n+1}}{6n+1}$$

(c) **Question:** Find the series representation for the solution of $\int \cos(e^x)\, dx$.

Solution:

$$\int \cos(e^x)\, dx = \sum_{n=0}^{\infty} ((-1)^n) \left(\frac{1}{2n!}\right) \frac{e^{2nx}}{2n} + C$$

Explanation: This is performed the same way. Since e^x is being sent in as the parameter for $\cos()$, we can replace every instance of x with e^x. Using Equation 25.10 as a base, when we substitute e^x in for x we get:

$$\cos(e^x) = \sum_{n=0}^{\infty} ((-1)^n) \left(\frac{1}{2n!}\right) e^{2nx}$$

Integrating $\cos(e^x)\, dx$ gives us:

$$\int \cos(e^x)\, dx = \sum_{n=0}^{\infty} ((-1)^n) \left(\frac{1}{2n!}\right) \frac{e^{2nx}}{2n} + C$$

Chapter 27

Thinking About Infinity

Label the order of infinity for each hyperreal number:

1. **Question:** $\omega^3 - \omega^2 + 32 + 0.1\omega^{-1}$

   Solution: 3

   Explanation: The highest power of infinity here is 3.

2. **Question:** $5.2 - 0.3\omega^{-1} + 200\omega^{-2}$

   Solution: 0

   Explanation: Real numbers are zero order infinities.

3. **Question:** $3.2\epsilon + 1.2\epsilon^2$

   Solution: -1

   Explanation: ϵ is merely ω^{-1}, so the highest order of infinity is -1.

Find the standard part of each number:

4. **Question:** $3\omega^2 + 5 - 52^{-1}$

   Solution: ∞

   Explanation: The standard part of a real number with a positive order is "real infinity," using the sign of the coefficient of the highest order infinity.

5. **Question:** $-3\omega^2 + 200\omega + 1000 + 20000\epsilon + 11100\omega^{-2}$

   Solution: $-\infty$

   Explanation: Because the order is positive, it is a real infinity. Because the coefficient of the highest infinity is negative, it is a negative infinity.

6. **Question:** $5 + 1.2\omega^{-2}$

   Solution: 5

   Explanation: The standard part of a number with order 0 is simply the real number at order zero.

7. **Question:** $\omega^{-5} - \omega^2$

   Solution: $-\infty$

   Explanation: This one was written backwards, but notice the highest order is two, and it doesn't have a coefficient (so the coefficient is one), but it is *subtracted*, so the coefficient is actually negative 1. Since this is a positive order, it will be real infinity. Since the coefficient is negative, it will be a negative infinity.

8. **Question:** $0.0035 + 100.3\epsilon$

   Solution: 0.0035

   Explanation: Because the order is zero, it is a real number, and the zero order coefficient is the standard part.

9. **Question:** $\epsilon + 2.34$

Solution: 2.34

Explanation: The order is zero, therefore it is a real number. The real number (the one without an infinity or infinitesimal) is the standard part.

Perform the following arithmetic. Find the hyperreal value. Then find the standard part.

10. **Question:** $\frac{\omega^2 - 3}{\omega^5}$

Solution: The hyperreal value is $\omega^{-3} - 3\omega^{-5}$ and the standard part is 0.

Explanation: Since the denominator is a single term, we can divide each part of the numerator by this term:

$$\frac{\omega^2}{\omega^5} - \frac{3}{\omega^5}$$

This simplifies to:

$$\omega^{-3} - 3\omega^{-5}$$

This is the hyperreal value.. Since the order is negative (-3), then the standard part of this number is zero.

11. **Question:** $(1.2\omega - 5)^2$

Solution: The hyperreal value is $1.44\omega^2 + -12\omega + 25$. The standard part is ∞.

Explanation: Remember, the ω acts a lot like a variable, whose value just happens to be really, really big. Therefore, we can expand this just like we did in algebra:

$$(1.2\omega - 5)^2 = (1.2\omega)^2 + 2(-5)(1.2\omega) + (-5)^2$$

$$= 1.44\omega^2 + -12\omega + 25$$

This is the hyperreal. Since the order is positive (2), that means that this is an infinity. Since

the coefficient of the highest order is positive, that means that it is a positive infinity, or ∞.

12. **Question:** $\frac{\sqrt{\omega^2 + 3\omega^2} + 6}{\omega}$

Solution: The hyperreal value is $\pm 2 + 6\omega^{-1}$. The standard part is ± 2.

Explanation: Again, remember to simplify this similar to an algebra problem. First, we will combine the terms under the square root:

$$\frac{\sqrt{4\omega^2} + 6}{\omega}$$

Now notice that the value under the square root is, well, square. Therefore, we can just pull it out (though we don't know if it is positive or negative):

$$\frac{\pm 2\omega + 6}{\omega}$$

Since the denominator is a single term, we can split the fraction as follows:

$$\frac{\pm 2\omega}{\omega} + \frac{6}{\omega}$$

This simplifies to:

$$\pm 2 + 6\omega^{-1}$$

This is the hyperreal value. Since this is a zero-order hyperreal, then the standard part is just the real number, which is ± 2.

Answer the following questions:

13. **Question:** Why do first order differentials generally produce a real value when put into ratio with each other.

Solution: Differentials are hyperreals with order -1. Since they are the same order, dividing them by each other causes the orders of infinity to cancel out.

14. **Question:** If an infinitesimal is added to a real number, what happens to the standard part of the number?

Solution: Nothing happens to the standard part of a real number when an infinitesimal is added.

Chapter 28

Limits: Finding Impossible and Non-Existent Values of Functions

1. **Question:** In the following graph, what is the limit as x approaches 2 from the left and the right? What is the actual value of the function at $x = 2$? Write the answer using limit notation.

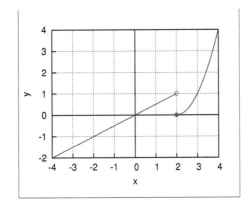

Solution:

$$\lim_{x \to 2^-} f(x) = 1$$

$$\lim_{x \to 2^+} f(x) = 0$$

$$f(x) = 0$$

Explanation: The left-hand limit is headed towards 1 and the right-hand limit is headed towards 0. The filled circle represents the actual value, so that is the value of the function.

Evaluate the following limits without using ϵ:

2. **Question:** What is $\lim_{x \to 5} 4x$?

Solution: 20

Explanation: For most values of most functions, the limit is simply the value of the function for the given x value.

3. **Question:** What is $\lim_{x \to 0} \frac{x^3}{x-1}$?

Solution: 0

Explanation: For most values of most functions, the limit is simply the value of the function for the given x value.

4. **Question:** What is $\lim_{x \to 5} \frac{x^2-25}{x-5}$?

Solution: 10

Explanation: To solve this we want to get rid of the denominator, so we will try factoring the numerator. The numerator can be factored into $(x + 5)(x - 5)$. Then $x - 5$ can be cancelled out of the denominator. This leaves the function as $x + 5$. Replacing x with 5 will yield 10.

5. **Question:** What is $\lim_{x \to 2} \frac{x^2-3x+2}{x-2}$?

Solution: 1

Explanation: To solve this we want to get rid of the denominator, so we will try factoring the numerator. The numerator can be factored into $(x - 2)(x - 1)$. Now $x - 2$ can be cancelled out from the denominator, leaving $x - 1$. At $x = 2$, this makes $2 - 1 = 1$.

Solve the following limits by adding in ϵ (remember to take both the right-hand and left-hand limits):

6. **Question:** Find $\lim\limits_{x \to 2} \frac{x^3 - 8}{6x - 12}$

Solution: 2

Explanation: In this problem, we will need to add ϵ to the x value to evaluate the right-hand limit:

$$\lim_{x \to 2^+} \frac{x^3 - 8}{6x - 12} = \frac{(2 + \epsilon)^3 - 8}{6(2 + \epsilon) - 12}$$

$$= \frac{2^3 + 3(2)^2\epsilon + 3(2)\epsilon^2 + \epsilon^3 - 8}{12 + 6\epsilon - 12}$$

$$= \frac{12\epsilon + 6\epsilon^2 + \epsilon^3}{6\epsilon}$$

$$= \frac{12\epsilon}{6\epsilon} + \frac{6\epsilon^2}{6\epsilon} + \frac{\epsilon^3}{6\epsilon}$$

$$= 2 + \epsilon + \frac{1}{6}\epsilon^2$$

This final value is a hyperreal with a standard part of 2.

Now for the left-hand limit:

$$\lim_{x \to 2^-} \frac{x^3 - 8}{6x - 12} = \frac{(2 - \epsilon)^3 - 8}{6(2 - \epsilon) - 12}$$

$$= \frac{2^3 - 3(2)^2\epsilon + 3(2)\epsilon^2 - \epsilon^3 - 8}{12 - 6\epsilon - 12}$$

$$= \frac{-12\epsilon + 6\epsilon^2 - \epsilon^3}{-6\epsilon}$$

$$= \frac{-12\epsilon}{-6\epsilon} + \frac{6\epsilon^2}{-6\epsilon} + \frac{-\epsilon^3}{-6\epsilon}$$

$$= 2 - \epsilon + \frac{1}{6}\epsilon^2$$

This final value is a hyperreal with a standard part of 2.

Therefore, since both limits agree, the limit is 2.

7. **Question:** Find $\lim\limits_{x \to -2} \frac{x^3 + 8}{x + 2}$

Solution: 12

Explanation: We will solve for the right-hand limit by adding ϵ to x:

$$\lim_{x \to -2^+} \frac{x^3 + 8}{x + 2} = \frac{(-2 + \epsilon)^3 + 8}{(-2 + \epsilon) + 2}$$

$$= \frac{-8 + 3 \cdot 4 \cdot \epsilon + 3 \cdot -2 \cdot \epsilon^2 + 8}{\epsilon}$$

$$= \frac{12\epsilon - 2\epsilon^2}{\epsilon}$$

$$= 12 - 2\epsilon$$

The standard part of this value is 12.

Now we will solve for the left-hand limit by subtracting ϵ:

$$\lim_{x \to -2^-} \frac{x^3 + 8}{x + 2} = \frac{(-2 - \epsilon)^3 + 8}{(-2 - \epsilon) + 2}$$

$$= \frac{-8 + 3(4)(-\epsilon) + 3(-2)(-\epsilon)^2 + 8}{-\epsilon}$$

$$= \frac{-12\epsilon - 6\epsilon^2}{-\epsilon}$$

$$= 12 + 2\epsilon$$

The standard part of this value is 12.

Since both limits agree, 12 is indeed the limit.

8. **Question:** Find $\lim\limits_{x \to 0} \frac{x}{x^2}$.

Solution:

$$\lim_{x \to 0^+} \frac{x}{x^2} = \infty$$

$$\lim_{x \to 0^-} \frac{x}{x^2} = -\infty$$

Explanation: To find the limit at zero, we will replace x with $0 + \epsilon$ on the right, and $x - \epsilon$ on the left. Let's start with the right-hand limit:

$$\lim_{x\to 0^+} \frac{x}{x^2} = \frac{0+\epsilon}{(0+\epsilon)^2}$$

$$= \frac{\epsilon}{\epsilon^2}$$

$$= \frac{1}{\epsilon}$$

$$= \epsilon^{-1}$$

$$= \omega$$

Remember that since ϵ is just ω^{-1}, if we get ϵ^{-1} then that will be the same as ω.

Therefore, the standard part for this value is an infinity. It will be a positive infinity because the coefficient of the highest-order term is positive.

Therefore, $\lim_{x\to 0^+} \frac{x}{x^2} = \infty$.

Now, let's do the same for the left-hand limit:

$$\lim_{x\to 0^-} \frac{x}{x^2} = \frac{0-\epsilon}{(0-\epsilon)^2}$$

$$= \frac{-\epsilon}{(-\epsilon)^2}$$

$$= \frac{-\epsilon}{\epsilon^2}$$

$$= \frac{-1}{\epsilon}$$

$$= -\epsilon^{-1}$$

$$= -\omega$$

The standard part for this value is also an infinity. It will be a negative infinity because the coefficient of the highest-order term is negative.

Therefore, $\lim_{x\to 0^-} \frac{x}{x^2} = -\infty$.

You can see this in the graph of the function:

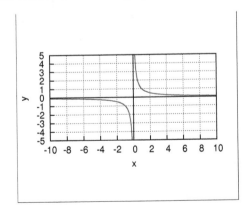

9. **Question:** Find $\lim_{x\to 1} \frac{(x-1)^2}{x-1}$.

Solution: 0

Explanation: To find the limit, we will start by taking the right-hand limit, by adding ϵ to x.

$$\lim_{x\to 1^+} \frac{(x-1)^2}{x-1} = \frac{((x+\epsilon)-1)^2}{(x+\epsilon)-1}$$

$$= \frac{x^2 + 2x\epsilon + \epsilon^2 - 2x + -2\epsilon + 1}{x+\epsilon-1}$$

$$= \frac{(1)^2 + 2(1)\epsilon + \epsilon^2 - 2(1) - 2\epsilon + 1}{1+\epsilon-1}$$

$$= \frac{1+\epsilon^2-1}{\epsilon}$$

$$= \frac{\epsilon^2}{\epsilon}$$

$$= \epsilon$$

Since the result was just ϵ, that means that the standard part is 0.

Now we need to do the left-hand limit:

$$\lim_{x\to 1^-} \frac{(x-1)^2}{x-1} = \frac{((x-\epsilon)-1)^2}{(x-\epsilon)-1}$$

$$= \frac{x^2 - 2x\epsilon + \epsilon^2 - 2x + 2\epsilon + 1}{x-\epsilon-1}$$

$$= \frac{(1)^2 - 2(1)\epsilon + \epsilon^2 - 2(1) + 2\epsilon + 1}{1-\epsilon-1}$$

$$= \frac{1+\epsilon^2-1}{-\epsilon}$$

$$= \frac{\epsilon^2}{-\epsilon}$$

$$= -\epsilon$$

The stanard part for this is 0 as well, so the final limit is zero.

Chapter 29

Limits: Finding Results Near Infinity

Evaluate the following limits:

1. **Question:** Find $\lim\limits_{x \to \infty} \frac{3x^5 - 2x^3 + 60x^2 - 300}{4x^5 + 5x^5 - 33x^3 + 1000x}$

   Solution: $\frac{3}{4}$

   Explanation: When dealing in limits to infinity, only the highest degree term from the numerator and the highest degree term from the denominator affect the outcome. As x becomes bigger and bigger, the lower degree terms become less and less significant in the long-term behavior. As x proceeds towards infinity, the lower degree terms become infinitely less significant.

   Therefore, by simplifying the numerator and the denominator to their highest degree, we can see that this reduces to:

   $$\frac{3x^5}{4x^5}$$

   The xs cancel, leaving:

   $$\frac{3}{4}$$

2. **Question:** Find $\lim\limits_{x \to \infty} \frac{5x^3 - 3x^2 + 600}{6x^4 - 500x + 10}$

   Solution: 0

   Explanation: When dealing in limits to infinity, only the highest degree term from the numerator and the highest degree term from the denominator affect the outcome. In the numerator, the highest degree term is $5x^3$. In the denominator, the highest degree term is $6x^4$.

Therefore, the problem reduces to:

$$\frac{5x^3}{6x^4}$$

Canceling the xs gives us:

$$\frac{5}{6x}$$

Notice that there is an x left in the denominator. Therefore, as x travels towards infinity, the denominator will become infinitely large. If you divide 5 into an infinite number of pieces, they will get smaller and smaller until they are indistinguishable from 0.

This can also be seen using hyperreals. If we substitute ω for x, we get $\frac{5}{6\omega}$, which we can write as either $\frac{5}{6}\omega^{-1}$ or $\frac{5}{6}\epsilon$. In other words, the result is an infinitesimal, so the standard part is 0.

3. **Question:** Find $\lim\limits_{x \to \infty} \frac{10x^7 + 600x^3 + 231x^2}{5x^7 - 20x^3 + 209}$

   Solution: 2

   Explanation: When dealing in limits to infinity, only the highest degree term from the numerator and the highest degree term from the denominator affect the outcome. In the numerator, the highest degree term is $10x^7$. In the denominator, the highest degree term is $5x^7$.

   Therefore, the problem reduces to:

   $$\frac{10x^7}{5x^7}$$

   Canceling the xs gives $\frac{10}{5}$ or 2.

4. **Question:** Find $\lim\limits_{x\to\infty} \frac{32x^6-65x^4+33}{88x^5+23x^4-25x^3+22x^2+100}$

Solution: ∞

Explanation: When dealing in limits to infinity, only the highest degree term from the numerator and the highest degree term from the denominator affect the outcome. In the numerator, the highest degree term is $32x^6$. In the denominator, the highest degree term is $88x^5$.

Therefore, the problem reduces to:

$$\frac{32x^6}{88x^5}$$

Canceling the xs gives us:

$$\frac{32x}{88}$$

Notice that there is an x left in the numerator. Therefore, as x travels towards infinity, this value will become infinitely large. Since we are traveling to positive infinity, the result will be a positive infinity.

This can also be seen using hyperreals. If we substitute ω for x, we get $\frac{32}{88}\omega$, which is a hyperreal infinity. The standard part is ∞.

5. **Question:** Find $\int_5^\infty \frac{7}{x^3}\,dx$

Solution: $\frac{21}{625}$

Explanation: To find an integral with infinite limits of integration, first perform the indefinite integral normally, and then use limits to determine any results at infinity.

$$\int_5^\infty \frac{7}{x^3}\,dx = \int \frac{7}{x^3}\,dx\,\Big|_5^\infty$$

$$= \int 7x^{-3}\,dx\,\Big|_5^\infty$$

$$= -21x^{-4}\,\Big|_5^\infty$$

$$= -21(\infty)^{-4} - -21(5)^{-4}$$

$$= \frac{-21}{\infty^4} + \frac{21}{625}$$

Here, the term $\frac{-21}{\infty^4}$ can be evaluated using limit techniques. Since the infinity was in the denominator, $\frac{-21}{\infty^4}$ reduced to zero. Therefore, we can complete this evaluation as follows:

$$\lim_{h\to\infty} \frac{-21}{h^4} + \frac{21}{625} = 0 + \frac{21}{625}$$

$$= \frac{21}{625}$$

6. **Question:** Find $\int_0^\infty x^2\,dx$

Solution: ∞

Explanation: To find an integral with infinite limits of integration, first perform the indefinite integral normally, and then use limits to determine any results at infinity.

$$\int_0^\infty x^2\,dx = \int x^2\,dx\,\Big|_0^\infty$$

$$= \frac{x^3}{3}\,\Big|_0^\infty$$

$$= \frac{(\infty)^3}{3} - \frac{(0)^3}{3}$$

$$= \frac{(\infty)^3}{3} - 0$$

$$= \frac{(\infty)^3}{3}$$

The ∞ tells us we need to use our limit rules to determine the answer. Since the ∞ is in the numerator, the term $\frac{\infty^3}{3}$ will grow infinitely large. Therefore, it iself will become an infinite term.

Therefore, the answer is ∞.

7. **Question:** Find $\int_{-\infty}^{-3} 2x^{-5}\,dx$

Solution: $-\frac{1}{162}$

Explanation: To find an integral with infinite limits of integration, first perform the indefinite integral normally, and then use limits to determine any results at infinity.

$$\int_{-\infty}^{-3} 2x^{-5}\,dx = \int 2x^{-5}\,dx \Big|_{-\infty}^{-3}$$

$$= \frac{2x^{-4}}{-4} \Big|_{-\infty}^{-3}$$

$$= -\frac{1}{2x^4} \Big|_{-\infty}^{-3}$$

$$= -\frac{1}{2(-3)^4} - -\frac{1}{2(-\infty)^4}$$

$$= -\frac{1}{162} + \frac{1}{2(-\infty)^4}$$

The presence of ∞ tells us that we need to use our limit rules to determine the answer. Since the infinity is in the denominator, the term $\frac{1}{2(-\infty)^4}$ will get infinitely close to zero. In other words:

$$\lim_{h \to -\infty} \frac{1}{2h^4} = 0$$

This reduces the expression to:

$$-\frac{1}{162} + 0$$

Or just $-\frac{1}{162}$.

8. **Question:** Find $\lim\limits_{x \to \infty} \frac{2x^3 - 5x^9 + 3x^2 + 500}{3x^5 - 2x^4 + 12x^2 + 8x^9 - 5000}$

Solution: $-\frac{5}{8}$

Explanation: When dealing in limits to infinity, only the highest degree term from the numerator and the highest degree term from the denominator affect the outcome. In this case, the highest degree is not in the first position—you have to actually look at the problem to find it.

On the numerator, the highest degree term is $-5x^9$. On the denominator it is $8x^9$. Therefore, as x approaches infinity, the other terms become infinitely less important. Therefore, this fraction reduces to:

$$\lim_{x \to \infty} \frac{-5x^9}{8x^9}$$

The xs cancel out, which just leaves $-\frac{5}{8}$

9. **Question:** Find $\lim\limits_{x \to -\infty} \frac{x^4 + x^5}{x^3 - 5x^4}$

Solution: ∞

Explanation: When dealing in limits to infinity, only the highest degree term from the numerator and the highest degree term from the denominator affect the outcome. Therefore, this fraction reduces to $\frac{x^5}{-5x^4}$. This reduces to $-\frac{x}{5}$. As x approaches $-\infty$, this will get bigger and bigger (because the negatives are cancelling each other out). Therefore, the result will be ∞.

10. **Question:** Find $\lim\limits_{x \to \infty} \frac{\sin(x)}{x}$

Solution: 0

Explanation: This limit is interesting because the numerator is oscillating. Think about the sine function—it goes back and forth *forever* between −1 and 1. It does not have a single stable value at all. However, the numerator has a stable *range*—it never goes above 1 or below −1. This will allow us to reason about it better.

Now thing about the denominator. As x goes to infinity, the denominator is getting bigger and bigger. Note that the numerator is *not* getting bigger and bigger. It isn't just sitting there, but it is always bounded within its box, from −1 to 1.

Therefore, since dividing *any* value in that range by ∞ has a limit of 0, that means that this function must have a limit of 0. Even though we don't know what *specific* value to assign for $\sin(x)$, we know that no matter which value in its range we were to assign it, the limit would be 0.

Chapter 30

Limits: Difficult Limits with L'Hospital

Find the following limits using L'Hospital's Rule:

1. **Question:** $\lim\limits_{x \to 4} \frac{x^2 - 16}{x - 4}$

   Solution: $\lim\limits_{x \to 4} \frac{x^2 - 16}{x - 4} = 8$

   Explanation: At $x = 4$, this evalutes to $\frac{(4)^2 - 16}{(4) - 4} = \frac{0}{0}$ so L'Hospital's rule applies. Therefore, we take the differential of the top and the bottom:

   $$\lim_{x \to 4} \frac{x^2 - 16}{x - 4} = \frac{d(x^2 - 16)}{d(x - 4)}$$
   $$= \frac{2x \, dx}{dx}$$
   $$= 2x$$
   $$= 2(4)$$
   $$= 8$$

2. **Question:** $\lim\limits_{x \to 0} \frac{\sin(x)}{x}$

   Solution: $\lim\limits_{x \to 0} \frac{\sin(x)}{x} = 1$

   Explanation: At $x = 0$, this evaluates to $\frac{\sin(0)}{0} = \frac{0}{0}$ so L'Hospital's rule applies. Therefore, we take the differential of the top and the bottom:

   $$\lim_{x \to 0} \frac{\sin(x)}{x} = \frac{d(\sin(x))}{d(x)}$$
   $$= \frac{\cos(x) \, dx}{dx}$$
   $$= \cos(x)$$
   $$= \cos(0)$$
   $$= 1$$

3. **Question:** $\lim\limits_{x \to 0} \frac{2x - \sin(x)}{x}$

   Solution: 1

   Explanation: At $x = 0$, this evaluates to $\frac{0 - 0}{0} = \frac{0}{0}$, which is an indeterminate form to which we can apply L'Hospital's rule. Therefore, we take the differential of the top and bottom. This yields:

   $$\lim_{x \to 0} \frac{2x - \sin(x)}{x} = \frac{d(2x - \sin(x))}{d(x)}$$
   $$= \frac{(2 - \sin(x)) dx}{dx}$$
   $$= 2 - \cos(x)$$
   $$= 2 - \cos(0)$$
   $$= 2 - 1$$
   $$= 1$$

4. **Question:** $\lim\limits_{x \to 1} \frac{x^2 - 1}{x^2 + 3x - 4}$

   Solution: $\lim\limits_{x \to 1} \frac{x^2 - 1}{x^2 + 3x - 4} = \frac{2}{5}$

   Explanation: At $x = 1$, this reduces to $\frac{(1)^2 - 1}{(1)^2 + 3(1) - 4} = \frac{0}{0}$, so L'Hospital's rule applies. Therefore, we take the differential of the top and bottom to find the result:

$$\lim_{x \to 1} \frac{x^2 - 1}{x^2 + 3x - 4} = \frac{d(x^2 - 1)}{d(x^2 + 3x - 4)}$$

$$= \frac{2x\,dx}{(2x + 3)\,dx}$$

$$= \frac{2x}{2x + 3}$$

$$= \frac{2(1)}{2(1) + 3}$$

$$= \frac{2}{5}$$

5. Question: $\lim_{x \to \infty} x\,e^{-2x}$

Solution: 0

Explanation: As x approaches infinity, this becomes the form $\infty \cdot 0$. To convert it to a form for L'Hospital's Rule, we can use the negative in the exponent and rewrite it in the form $\frac{x}{e^{2x}}$. This form is $\frac{\infty}{\infty}$, for which we can apply L'Hospital's Rule. This becomes:

$$\lim_{x \to \infty} \frac{x}{e^{2x}} = \frac{d(x)}{d(e^{2x})}$$

$$= \frac{dx}{2e^{2x}\,dx}$$

$$= \frac{1}{2e^{2x}}$$

As x approaches infinity, this expression approaches zero.

6. Question: $\lim_{x \to 0} \frac{\sin(x^2)}{x \tan(x)}$

Solution: $\lim_{x \to 0} \frac{\sin(x^2)}{x \tan(x)} = 1$

Explanation: At $x = 0$, this reduces to $\frac{\sin((0)^2)}{(0)\tan(0)} = \frac{0}{0}$, so L'Hospital's rule applies. Therefore, we take the differential of the top

and bottom to find the result:

$$\lim_{x \to 0} \frac{\sin(x^2)}{x \tan(x)} = \frac{d(\sin(x^2))}{d(x \tan(x))}$$

$$= \frac{2x \cos(x^2)\,dx}{(x \sec^2(x) + \tan(x))\,dx}$$

$$= \frac{2x \cos(x^2)}{x \sec^2(x) + \tan(x)}$$

$$= \frac{2(0) \cos((0)^2)}{(0) \sec^2(0) + \tan(0)}$$

$$= \frac{0}{0}$$

In this case, the result is still $\frac{0}{0}$! This means that we need to apply L'Hospital's rule again.

$$\lim_{x \to 0} \frac{2x \cos(x^2)}{x \sec^2(x) + \tan(x)}$$

$$= \frac{d(2x \cos(x^2))}{d(x \sec^2(x) + \tan(x))}$$

$$= \frac{(-4x^2 \sin(x^2) + 2 \cos(x^2))\,dx}{(2x \sec^2(x) \tan(x) + \sec^2(x) + \sec^2(x))\,dx}$$

$$= \frac{-4x^2 \sin(x^2) + 2 \cos(x^2)}{2x \sec^2(x) \tan(x) + 2 \sec^2(x)}$$

$$= \frac{-4(0)^2 \sin((0)^2) + 2 \cos((0)^2)}{2(0) \sec^2(0) \tan(0) + 2 \sec^2(0)}$$

$$= \frac{0 + 2}{0 + 2}$$

$$= \frac{2}{2}$$

$$= 1$$

7. Question: $\lim_{x \to \infty} \frac{e^{3x}}{4x + 200}$

Solution: $\lim_{x \to \infty} \frac{e^{3x}}{4x + 200} = \infty$

Explanation: First, to evaluate this at infinity, we need to use the hyperreal unit ω. This yeilds $\frac{e^{3\omega}}{4\omega + 200}$, which is an infinity over another infinity. Therefore, L'Hospital's rule applies. Applying L'Hospital's rule yields:

$$\lim_{x\to\infty} \frac{e^{3x}}{4x+200} = \frac{d(e^{3x})}{d(4x+200)}$$

$$= \frac{3e^{3x}\,dx}{4\,dx}$$

$$= \frac{3e^{3x}}{4}$$

$$= \frac{3}{4}e^{3x}$$

$$= \frac{3}{4}e^{3\omega}$$

In this case, the highest order term has a positive infinity, so the limit is ∞.

8. **Question:** $\lim_{x\to\infty} \frac{3x+2^x}{2x+3^x}$

Solution: $\lim_{x\to\infty} \frac{3x+2^x}{2x+3^x} = 0$

Explanation: This limit can be evaluated using ω as the hyperreal unit of infinity. Using this substitution yields $\frac{3\omega+2^\omega}{2\omega+3^\omega}$, but not in a way that is easily reducible. Both of these are infinite terms, so this is another instance of $\frac{\infty}{\infty}$. Therefore, L'Hospital's rule applies, and we can use it to simplify things further. Therefore, we can apply the rule:

$$\lim_{x\to\infty} \frac{3x+2^x}{2x+3^x} = \frac{d(3x+2^x)}{d(2x+3^x)}$$

$$= \frac{(3+\ln(2)\,2^x)\,dx}{(2+\ln(3)\,3^x)\,dx}$$

$$= \frac{3+\ln(2)\,2^x}{2+\ln(3)\,3^x}$$

This will also reduce to $\frac{\infty}{\infty}$, so we can perform L'Hospital's rule again:

$$\lim_{x\to\infty} \frac{3+\ln(2)\,2^x}{2+\ln(3)\,3^x} = \frac{d(3+\ln(2)\,2^x)}{d(2+\ln(3)\,3^x)} = \frac{0+\ln(2)\,\ln(2)\,2^x}{0+\ln(3)\,\ln(3)\,3^x}$$

$$= \frac{\ln(2)^2\,2^x}{\ln(3)^2\,3^x}$$

Since, in both fractions, both the numerator and the denominator have the same exponent, we can move the exponent outside of the fraction. This gives:

$$\lim_{x\to\infty} \frac{\ln(2)^2\,2^x}{\ln(3)^2\,3^x} = \left(\frac{\ln(2)}{\ln(3)}\right)^2 \left(\frac{2}{3}\right)^x$$

$$= \left(\frac{\ln(2)}{\ln(3)}\right)^2 \left(\frac{2}{3}\right)^\omega$$

In this last formula, we have a number that is greater than 0 but less than 1 raised to an infinite power. If you think about it, if you have a number greater than 0 but less than 1, every time you multiply it by itself it gets smaller (closer to zero). So, if you do that an infinite number of times, it will get infinitely close to zero. Therefore, in the limit, it reduces down to zero! This gives us:

$$\lim_{x\to\infty} \left(\frac{\ln(2)}{\ln(3)}\right)^2 \left(\frac{2}{3}\right)^x = \left(\frac{\ln(2)}{\ln(3)}\right)^2 \left(\frac{2}{3}\right)^\omega$$

$$= \left(\frac{\ln(2)}{\ln(3)}\right)^2 \cdot 0$$

$$= 0$$

9. **Question:** $\lim_{x\to 5^+} \frac{x}{x-5}$

Solution: $\lim_{x\to 5^+} \frac{x}{x-5} = \infty$

Explanation: If you evaluate this at $x=5$, you get $\frac{5}{5-5}$. This is $\frac{5}{0}$. This is *not* a form that L'Hospital's rule supports, so we *cannot* differentiate the top and the bottom. Instead, we can use $5+\epsilon$ to see what this should resolve to. This becomes:

$$\lim_{x\to 5^+} \frac{x}{x-5} = \frac{5+\epsilon}{5+\epsilon-5}$$

$$= \frac{5+\epsilon}{\epsilon}$$

$$= \frac{5}{\epsilon} + \frac{\epsilon}{\epsilon}$$

$$= 5\omega + 1$$

Since this is an infinite value, the limit is ∞.

10. **Question:** $\lim_{x\to\infty} (1+\frac{1}{x})^x$

Solution: $\lim_{x\to\infty} (1+\frac{1}{x})^x = e$

Explanation: If you substitute ω in for x, this yields 1^∞. This is one of the forms that L'Hospital's rule applies to, but we have to manipulate it to get it in a form to use L'Hospital's rule. Since it uses exponents, we will have to remove the exponent through logarithms. We will start by setting the limit to a value C:

$$C = (1 + \frac{1}{x})^x$$

$$\ln(C) = \ln\left((1 + \frac{1}{x})^x\right)$$

$$\ln(C) = x \ln(1 + \frac{1}{x})$$

As x heads towards infinity $\ln(1 + \frac{1}{x})$ heads to zero. However, since we are multiplying this by x as well, this is now of the form $\infty \cdot 0$. Therefore, we can transform x into $\frac{1}{\frac{1}{x}}$ to change it into a $\frac{0}{0}$ form. Therefore, we have:

$$\ln(C) = \frac{1}{\frac{1}{x}} \ln(1 + \frac{1}{x})$$

$$= \frac{\ln(1 + \frac{1}{x})}{\frac{1}{x}}$$

$$= \frac{d(\ln(1 + \frac{1}{x}))}{d\left(\frac{1}{x}\right)}$$

$$= \frac{\frac{-x^{-2}\,dx}{1 + \frac{1}{x}}}{-x^{-2}\,dx}$$

$$= \frac{1}{1 + \frac{1}{x}}$$

$$= \frac{1}{1 + \frac{1}{\omega}}$$

$$= \frac{1}{1}$$

$$\ln(C) = 1$$

$$e^{\ln(C)} = e^1$$

$$C = e$$

Therefore, the limit of this is e.

11. **Question:** $\lim\limits_{x \to \infty} (2 + \frac{1}{3x})^x$

Solution: ∞

Explanation: If you substitute ω in for x, this yields 2^ω, which is just ∞. This is not a form for L'Hospital's rule, so the result is just ∞.

Chapter 31

Conclusion

This chapter had no questions.

CPSIA information can be obtained
at www.ICGtesting.com
Printed in the USA
LVHW021629170723
752705LV00043B/857

9 781944 918156